Bats of Puerto Rico

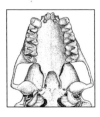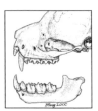

B   o

and a Caribbean Perspective

Michael R. Gannon
Allen Kurta
Armando Rodríguez-Durán
Michael R. Willig
Illustrations by Jeffrey Martz

The University of the West Indies Press

Jamaica • Barbados • Trinidad and Tobago

This book is typeset in Adobe Garamond. The paper used in this book meets the minimum requirements of ANSI/NISO Z39.48-1992 (R1997). ∞

A catalogue record of this book is available from the National Library of Jamaica.

Printed in the United States of America

05 06 07 08 09 10 11 12 13 / 9 8 7 6 5 4 3 2 1

ISBN 976-640-175-6

The University of the West Indies Press
1A Aqueduct Flats, Mona
Kingston 7, Jamaica
www.uwipress.com

*To my parents, Robert and Rose, for their never failing
inspiration and encouragement.*
MICHAEL R. GANNON

*To my graduate students, for ultimately becoming friends and
family; to Mark, for sharing my adventures.*
ALLEN KURTA

*To my father, who taught me the love for knowledge and the
appreciation of nature; to my mother, who planted the seed of skepticism.*
ARMANDO RODRÍGUEZ-DURÁN

*To my teachers, who planted the seeds of knowledge; to my
colleagues, who nurtured understanding; to my students, who cultivated
wisdom.*
MICHAEL R. WILLIG

Contents

Preface

Bats have always fascinated humans. They are the only true flying mammals, and they are active mostly at night, making them appear secretive and mysterious. Although historically revered in many Eastern cultures, they are among the most misunderstood animals in the modern Western world, forming the basis for many fictional tales and stories of evil and the supernatural. From ecological and economic perspectives, however, bats are incredibly important. They are major nocturnal predators of insects and play essential roles in pollinating plants and dispersing seeds. Few people realize the impact that bats have on our everyday lives or the many plants and plant products that are dependent on bats.

Bats are particularly significant on tropical islands, where they often dominate the native mammalian fauna in number of species and number of individuals. This is true, for example, on most islands of the Caribbean, and Puerto Rico is no exception, harboring only 13 living species of native mammals, all of which are bats. Although the number of species on an island typically is less than in a comparable area on the mainland, islands often harbor endemic species or genera that have evolved in isolation from their mainland ancestors. For instance, six of the 13 species of bat that live on Puerto Rico occur only on that land mass and other nearby islands.

Our goals are to provide an introduction to the Puerto Rican ecosystem, placing it within the appropriate geographic, historical, and cultural context, and to synthesize available information on the ecology, behavior, and natural history of the bats of Puerto Rico. To do so, we relied on the published literature and incorporated many of our own unpublished observations and those of our students. We based our story on observations made on Puerto Rico, but when such information was lacking, we incorporated data from elsewhere in the Caribbean and occasionally the mainland to describe the life history of a species. In particular, we often consulted the excellent and detailed work of Gilberto Silva-Taboada (1979)—*Los Murciélagos de Cuba*—the only other book that focuses on bats of the Antilles.

Our book is intended for travelers, naturalists, teachers, and wildlife technicians as well as research biologists. We tried to minimize use of specialized jargon and to define technical terms in the text or glossary so that the book would be comprehensible to nonspecialists with a basic, undergraduate background in biology. To improve readability, we minimized use of scientific names (except for parasites, which often lack common names) and avoided long lists of citations in the text. Much technical information, of value primarily to specialists, is summarized in the appendices.

This book is a result of nearly 90 years of combined experience among the authors in the study of bats and 60 years specifically concerning bats of Puerto Rico. We hope it conveys the admiration we hold for these amazing creatures and our respect for their island home. In addition, we hope it will interest a new generation of biologists to continue the study of bats on Puerto Rico and other islands of the Caribbean.

Organization

The bulk of the text consists of three parts: a description of Puerto Rican ecosystems, a primer on the biology of bats, and a detailed description of the species that inhabit Puerto Rico. We begin by summarizing the geological history of the island and detailing aspects of its current geography. Next we describe physical factors, such as temperature and rainfall, and indicate how geography and climate interact to produce various types of forest and the general way in which humans have modified the forests of Puerto Rico. The first section concludes with a summary of the living and extinct mammals of Puerto Rico and a discussion of how Puerto Rico compares with other islands of the Antilles in terms of the number of species of mammals and bats that are present.

"Lack of information is one of the greatest, but perhaps least appreciated, threats to bats" (Hutson et al., 2001:41), and consequently, the second part of the text describes many of the intriguing aspects of the biology of bats—the combination of unique features that sets these animals apart from most other mammals. We describe some modifications of the body for flight, illustrate the basics of echolocation, and depict the variety of roosting sites occupied by these mammals, including caves and rock crevices, tree hollows and stumps, and even leaf tents and burrows within fallen trees. The feeding habits of bats make them "keystone taxa" in many tropical ecosystems, and we explain their roles as pollinators, seed dispersers, and insect predators and point out their essential involvement with little-known ecosystems within the lightless environments of caves. The limited potential for bats to be a source of disease for humans is discussed, and although most bats on Puerto Rico live in caves or trees, we conclude this section by giving advice on how to remove bats from buildings.

The main body of the book consists of accounts of the five families and 13 species of bat that live on Puerto Rico. The order in which families are presented follows Koopman (1993), and within families, species are listed in alphabetical order. Each species account contains a brief description of the animal's technical name, the derivation of its name, and often interesting information concerning the history of the species. Scientific names of mammals generally are those accepted by Wilson and Reeder (2005), whereas scientific names of plants usually follow Liogier and Martorell (1982). The common names we use for animals and plants generally are those that most frequently appear in the literature and/or are used most often by workers on the island. Although Wilson and Cole (2000) recently attempted to standardize common names for all extant mammals, including bats, we occasionally chose not to follow their suggestions because some of their proposed names are not in common use by biologists or lay persons. We do, however, indicate the names proposed by Wilson and Cole (2000) in the text and/or in one of the appendices when they differ from those used in this book.

In addition to information concerning the name of each Puerto Rican species, species accounts also contain succinct descriptions of the geographic range of each bat, illustrated with appropriate maps of Puerto Rico and the Caribbean basin. After providing a physical description of each species, including a representative portrait and skull drawing, we summarize the natural history of the species, including the particulars on roosting sites, activity patterns, diet, reproduction, and parasites.

Subsequent to the species accounts, we present recommendations for the conservation of bats; these are applicable to other islands of the Caribbean as well as Puerto Rico. We also provide keys for the identification of Puerto Rican bats based on an animal in hand or a preserved skull. The keys are followed by appendices tabulating various types of plants that provide food for Puerto Rican bats; the external body measurements, cranial measurements, ectoparasites, and chromosomal characteristics of each species; technical names of organisms mentioned in the text; and capture localities for each species of bat on Puerto Rico and nearby islands (i.e., a gazetteer). The book concludes with a detailed glossary and a list of references.

Acknowledgments

Comments made by G. R. Horst and T. H. Kunz greatly improved the manuscript. We thank Inter American University for partly subsidizing publication of this book and for providing ARD with time to work on it. ARD also acknowledges the following for assistance or for providing unpublished information: A. Brooke, E. Durán, R. Gonzalez, M. Quiles, B. Rivera-Marchand, I. Rivera, A. Rodríguez, C. Rodríguez, A. Ruiz, E. Santiago, J. Soto-Centeno, F. Torres, A. Vale, B. Valentín, and R. Vázquez as well as the Sociedad Avance Espeleológico. AK thanks N. Simmons for providing an advanced copy of her taxonomic summary of the Chiroptera; J. Kittel, K. Nichols, and especially J. Alexander for secretarial support; G. Hannan for botanical advice; M. Kurta and J. Soto-Centeno for assistance in the field; and the Office of Graduate Studies and Research, Eastern Michigan University, for financial support during fieldwork.

The cost of publishing this book also was defrayed partly through a grant to MRW from the U.S.D.A. Forest Service, International Institute for Tropical Forestry, through the aegis of A. Lugo and J. Wunderle. In addition, much of the research regarding the distributional records of Puerto Rican bats resulted from a series of projects directed by MRW and MRG and supported by the U.S. Department of Interior, Fish and Wildlife Service, through the cooperation of J. Saliva. MRW and MRG gratefully acknowledge a series of grants from the National Science Foundation (i.e., Long-term Ecological Research in the Luquillo Mountains) and from Oak Ridge Associated Universities; together these grants supported many years of fieldwork in the Luquillo Experimental Forest.

MRW is indebted to Texas Tech University for supporting fieldwork for over 19 years and providing a conducive "home environment" for distant tropical research. MRG thanks The Pennsylvania State University for supporting his fieldwork over the last 10 years and making much of his contribution to this book possible. For assistance in the field, MRW and MRG also acknowledge the cooperation and dedication of many undergraduate and graduate students at Texas Tech University and Pennsylvania State University. Although all cannot be mentioned individually, MRW and MRG especially are indebted to R. D. Stevens, C. Dick, S. J. Presley, and B. T. Croyle for help in surveying the bats of Puerto Rico.

Bats of Puerto Rico

Puerto Rican Ecosystems

Although Puerto Rico is an island, it harbors a complex and diverse suite of plants and animals, representing a dynamic balance between factors that enhance and those that diminish diversity. The tropical climate, elevational relief, and strategic location of Puerto Rico, in the midst of the islands of the Caribbean, favor an increase in the variety of organisms, whereas the island's small size, distance from the mainland, and history of disturbance tend to reduce diversity. Interactions among these factors have created a unique biota on Puerto Rico, distinct from that of any other island or nation.

Equally important, the various factors that shaped the evolutionary history of Puerto Rico created an island that is a model for understanding how historical, geographic, and environmental factors interact to affect the diversity and complexity of resident species. The story of Puerto Rico is singular, but at the same time, it contains major themes that illuminate the evolutionary theater and ecological play that transpire on all islands in the Caribbean as well as vast areas of the mainland Neotropics.

Geography

Puerto Rico lies at a fulcrum (18°35'–17°55' N and 67°15'–65°35' W) between the Lesser and Greater Antilles and is the easternmost extension of the latter (Map 1). It is about 2,500 kilometers southeast of New York City and 1,600 kilometers east-southeast of Miami. The fourth largest island in the Caribbean, it has a land area of approximately 8,900 kilometers². Its greatest length is slightly more than 170 kilometers, between Punta Higüero in the west and Punta Puerca in the east, whereas the greatest width is slightly less than 65 kilometers, between Punta Colón in the north and Punta Isabela in the south. Many smaller islands and cays are dispersed along its coastline, especially in the south and east. Vieques and Culebra are the largest satellite islands and lie to the east of Puerto Rico, whereas the sizable islands of Mona, Monito, and Desecheo are situated to the west.

Puerto Rico is the smallest of the Greater Antilles, with Cuba, Hispaniola, and Jamaica each comprising appreciably more area (114,525; 76,190; and 11,425 kilometers², respectively). Aside from these larger westerly islands, however, Puerto Rico is far larger than any other island in the Caribbean. It exceeds the total area of all land in a variety of island groups, such as the Virgin Islands (971 kilometers²), Windward Islands (837 kilometers²), Leeward Islands (5,245 kilometers²), Barbados (427 kilometers²), and continental islands off the northern coast of South America (6,845 kilometers²).

The topographic relief of Puerto Rico is dramatic, rising from sea level to over 1,200 meters in only 30 kilometers (Map 2a). Indeed, the Puerto Rican landscape is dominated by hills and low mountains, with only about 25 percent of the land considered level (Picó, 1974). Christopher Columbus purportedly described Puerto Rico to the queen of Spain by crumbling a piece of parchment to emphasize the impressive topographic relief of the island. About 55 percent of its land is below 150 meters in elevation, 21 percent is between 150 and 300 meters, and 24 percent lies above 300 meters (Wadsworth, 1949).

Modern geographers divide the island into three major physiographic regions: a zone of coastal plains, a northwestern karst region, and a mountainous interior (Mattson et al., 1990; Picó, 1974). The interior is dominated by a mountainous spine that acts as an insular divide, separating rainwaters that flow north from those that drain southward across the island. The spine begins at the west coast near Mayagüez, where the Urayoan Mountains and the Cerro de las Mesas converge (Map 2a). From there, the mountains extend in an almost uninterrupted fashion to the east, forming the Cordillera Central, which terminates at Aibonito. Some of the highest

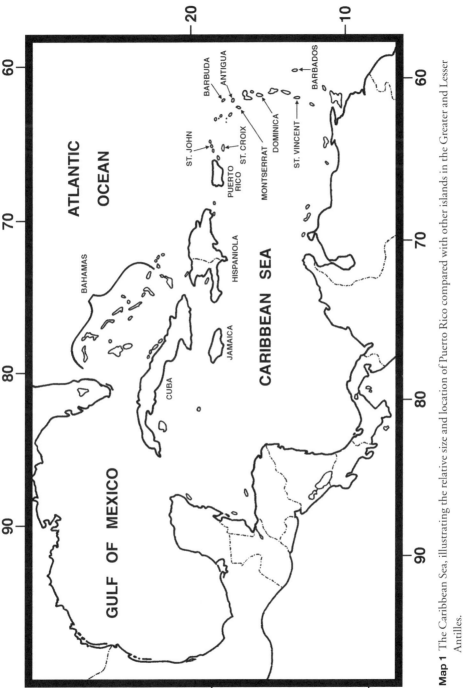

Map 1 The Caribbean Sea, illustrating the relative size and location of Puerto Rico compared with other islands in the Greater and Lesser Antilles.

peaks in Puerto Rico are in the Cordillera Central, such as Cerro de Punta (1,338 meters) and Cerro de Doña Juana (1,079 meters). The mountainous spine continues eastward between Salinas and Cayey along the Sierra de Cayey, where elevation diminishes to only 120 meters at its terminus between Las Piedras and Humacao. Nevertheless, extensions of the Sierra de Cayey—the sierras Guardarraya and Pandura—project northward and rise rapidly to form the Sierra de Luquillo, where elevations peak at 1,074 meters at El Toro (the bull) and 1,065 meters at El Yunque (the anvil).

Coastal areas are quite variable and consist of rocky outcrops and sand dunes, marshes and ponds, low rolling hills, alluvial fans, and low piedmont. Geographers identify seven major regions within the littoral areas of Puerto Rico. Among these are the northern coastal plain from the lower valley of the Río Grande de Arecibo to Cape San Juan (see Map 19 in Appendix 8 for locations of rivers); the narrow triangular valleys of the east (Fajardo, Naguabo, Antón Ruiz, Humacao, Yabucoa, and Maunabo valleys); and the southern coastal plain between Patillas and Ponce, including large alluvial fans between Guayama and Salinas (see Map 20 in Appendix 8 for locations of municipalities). Other geographic regions are the coastal valleys of the southwest, including the Ponce-Patillas coastal plain and flatlands near the Tallaboa, Guayanilla, and Yauco rivers; the Lajas Plains from Yauco and Guánica to Boquerón; and the southwestern hills, including the Sierra Bermeja (300 meters), which extend southwest from Guánica. The final coastal region consists of five river valleys in the west, including three large valleys (those of the Culebrinas, Añasco, and Guanajibo rivers) and two smaller ones (the Córcega and Yagüez rivers).

The northwestern karst region of Puerto Rico includes the Florida Hills and a low plateau known as the Mesita del Noroeste. Although the land is mostly flat near the coast, the topography becomes quite irregular toward the interior of the plateau. The region becomes so rugged "that many areas are entirely uninhabited, without even roads or trails crossing them; an exception indeed for densely populated Puerto Rico" (Picó, 1950:147). The area is underlain by limestone that slowly dissolves in the presence of even weak acids, such as the carbonic acid that forms when carbon dioxide in soil dissolves in rainwater (Lugo et al., 2001). Over centuries, differential dissolution of the limestone by this slightly acidified water results in a landscape punctuated by hundreds of cone-shaped hummocks or haystacks (*mogotes* or *pepinos*; Fig. 1, Color Plates), sinkholes (*sumideros*), and caves. Drainage from these areas creates a complex of subterranean rivers, one of which is the third largest underground river in the world—the Río Camuy.

Geological History

The geological origins of Puerto Rico are complex (Larue, 1994; Mattson et al., 1990), but the history of the island can be classified conveniently into five periods (Picó, 1974). The first period was a volcanic stage. Near the boundary of the Jurassic and Cretaceous periods, massive quantities of lava, gases, ashes, and powder emerged from fissures in the floor of the Atlantic Ocean and formed submarine mountains, which eventually became volcanic islands. Subsequently, surface erosion produced sediments, and reefs formed along the peripheries of the newly formed islands.

The second period occurred at the end of the Cretaceous and is known as the Antillean Revolution. Rocks that formed during the Cretaceous Period were compressed and folded into a chain of mountains that may have extended unbroken from present-day Cuba to the Virgin Islands. The folding pushed some rocks deep into the earth's crust, where they were melted again. The resulting magma eventually was forced upward, into the interior of the mountains, where it slowly cooled to form a variety of granitoid rocks, like those seen today between San Lorenzo and Yabucoa and in the valley of the Río Caonillas.

The third stage was primarily a time of erosion and limestone deposition. Many of the original mountains of Puerto Rico were eroded to form a peneplane (the St. John Peneplane) during the Lower Oligocene Epoch. However, the highest peaks of the Cordillera Central, Sierra de Cayey, and Sierra de Luquillo consisted of harder rocks that resisted the erosive forces of the time and remained as monadnocks above the surrounding peneplane. Between the Upper Oligocene and Lower Miocene epochs, the northern and southern fringes of the island became submerged and then covered by immense quantities of limestone sediment. In the north, these sediments persist today as sedimentary rocks occurring from Aguada to Loíza, although they are thickest in the karst area between Lares and Camuy. In the south, limestone deposits extended from present-day Cabo Rojo to Santa Isabel.

The fourth period was characterized by uplift and erosion. Some time after the Lower Miocene Epoch, Puerto Rico again was subjected to mountain-building forces that uplifted the entire island by 240 meters. Subsequent erosion of the raised area demolished the perimeter of the St. John Peneplane and heralded development of a new structure, the Caguana Peneplane.

Events during the fifth stage in the geological history of Puerto Rico were related to faulting, elevation, and even glaciation. Much of Puerto Rico's rectangular outline is a result of faulting that occurred along the edges of the island at the end of the Pliocene Epoch; rocks on the outer sides of the faults fell into the ocean, while the Caguana Peneplane was raised. Eventually, the borders of the peneplane were eroded to form coastal plains and valleys, and even some of the softer granitoid rocks of the interior were destroyed by these processes, ultimately producing the present-day plain that surrounds Caguas. Fluctuating sea level, caused by alternating formation and melting of continental glaciers during the Pleistocene Epoch, resulted in periods of submersion for parts of the island, and at these times, alluvium was deposited in coastal plains and valleys, eventually providing some of the soils that are now farmed. Today, the slow processes of uplift and erosion are continuing.

Temperature and Rainfall

Many factors affect the climate of Puerto Rico, including latitude, insularity, topography, and location with respect to large masses of land or water, trade winds, easterly waves, major air fronts, and hurricane routes. Compared with more temperate regions, the climate of Puerto Rico is consistently warm and wet. There is a marked period with reduced precipitation in winter and a period in summer when tropical storms and hurricanes are most common.

In general, the island has two zones of temperature defined by elevation. Hot, tropical regions characterize the lowlands and plains, whereas moderate, subtropical areas characterize the upper reaches of mountains. The 23°C (73°F) isotherm, which is more or less coincident with an elevation of 300 meters (Map 2b), defines the boundary between these zones. Mean annual temperature ranges from almost 27°C in coastal areas to 20°C at

ELEVATION CONTOURS

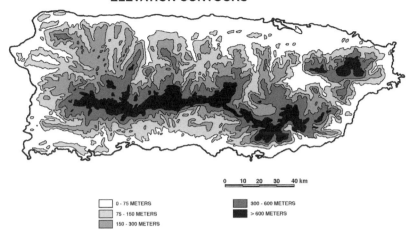

0 - 75 METERS	300 - 600 METERS
75 - 150 METERS	> 600 METERS
150 - 300 METERS	

TEMPERATURE CONTOURS (°F)

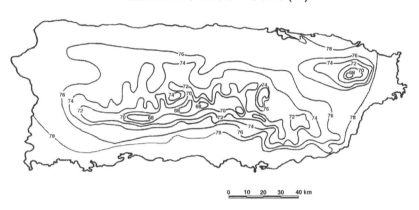

PRECIPITATION CONTOURS (INCHES/YEAR)

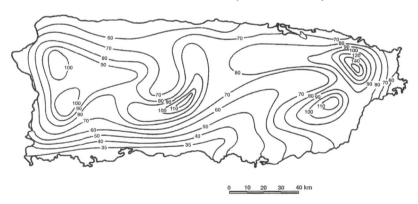

Map 2 Patterns of (a) elevation, (b) mean monthly temperature, and (c) mean monthly precipitation for Puerto Rico (Birdsey and Weaver, 1982; Picó, 1974).

mountain summits. August is usually the hottest month, and February is the coolest, although variation in temperature is small at any one location. For example, the difference between mean monthly temperatures for the hottest and coolest periods anywhere on the island is about 3–4°C. In contrast, average diurnal variation in temperature is about 11°C, although it is generally greater at high elevations (e.g., 13°C at Aibonito) and lower in coastal areas (e.g., 6°C at San Juan).

Puerto Rico, like much of the tropics, experiences large quantities of precipitation caused by either convective processes or the cooling effects of mountains (Map 2c). On the windward side of mountains, moisture-laden air derived from the trade winds is pushed upward and eventually cools, with the resulting condensation appearing as rain on the uplands and adjacent lowlands. The leeward side of mountains, however, normally receives little or no precipitation from this process and experiences a rain shadow.

Convective rain in contrast is a result of the high ambient temperatures typical of tropical regions. When the ground is hot, especially during times of maximum solar radiation (i.e., midday), air near the land is heated, grows lighter, and rises, before ultimately expanding and cooling at higher altitudes. As the air rises massive pillars of cloud are formed, and rain eventually falls, often as torrential showers of short duration.

Although geographic variation exists (Map 2a), precipitation exhibits a seasonal pattern; it is lowest for the first four or five months of the year, when the air temperature is lowest, and quite rainy at all other times. Peaks in rainfall vary during the year, depending on location. Most sites show two peaks, but the west coast has only one, and the Sierra de Luquillo may have three. Coastal areas generally receive less precipitation than do mountainous regions of the interior, and the northern coast receives more rainfall (150–180 centimeters annually) than does the southern coast (100–130 centimeters), with the area around Guánica being driest (75 centimeters). Annual precipitation is quite variable in the interior as well. Depending on elevation, yearly rainfall is between 150 and 250 centimeters in the Cordillera Central and in the Sierra de Cayey, whereas it ranges from 200 to 500 centimeters in the Sierra de Luquillo, the wettest location on the island.

Hurricanes

Hurricanes are common components of the climatic regime of Puerto Rico in particular and the Caribbean in general. They usually occur in the warmest months (June–November, with greatest frequency in September), striking the eastern or southern end of the island and proceeding west or northwest, following the general path of the trade winds (Picó, 1974). Most hurricanes originate west of the Cape Verde Islands in the southern North Atlantic, and those that impact Puerto Rico often form as wide low-pressure areas, which contract to form cyclonic whirlwinds that grow in intensity as they pass over warm waters of the Atlantic. Large hurricanes have diameters between 325 and 500 kilometers, whereas small hurricanes rarely exceed 80–160 kilometers in diameter. Although hurricanes vary considerably in size, their potential destructiveness is more closely related to how low barometric pressure becomes in the center (vortex) of the storm than to overall size.

Some damage may be wrought by a hurricane that merely approaches Puerto Rico, but the most destructive storms typically are those during which the vortex strikes the island. Between 1893 and 1956, vortices of six hurricanes (San Roque, San Ciriaco, San Felipe, San Nicolás, San Ciprián, and Santa Clara [Betsy]) passed directly over Puerto Rico, causing up to 3,000 human deaths and $50,000,000 in damage. Since 1956, the frequency of major hurricanes has diminished considerably, although there was a recent resurgence of activity associated with hurricanes Hugo (1989) and Georges (1998).

Beyond the misery and suffering imposed on humans, the high winds and immense rainfall associated with hurricanes alter both the biotic and abiotic portions of the environment in a variety of complex, cascading ways (Scatena and Larsen, 1991; Zimmerman et al., 1996; Waide, 1991). For example, rainfall accompanying a hurricane alters the physical environment by initiating landslides and floods (Fig. 2, Color Plates; Walker, 1991; Walker et al., 1996), whereas wind destroys or damages mature trees in the forest canopy. Reduction of the forest canopy, in turn, leads to increased levels of temperature and light below the canopy as well as increased amounts of forest litter from the dead and injured trees (Fig. 3a and 3b, Color Plates; Brokaw and Grear, 1991; Fernández and Fetcher, 1991).

Damage to mature trees suppresses reproduction by these plants for months or years following a hurricane, and even if seedlings of canopy trees survive the storm, many are injured by sunburn or buried under the rain of debris from dead and dying plants to the point that the young plants die from lack of sunlight (You and Petty, 1991). Seedlings from pioneer species, in contrast, often are favored by the overall increase in light intensity (Brokaw and Grear, 1991; Fernández and Fetcher, 1991).

Animals, of course, potentially are affected by storm-related modifications of their environment as well (Willig and McGinley, 1999). For example, the coquí, a common tree frog, experienced an increase in adult survivorship after Hurricane Hugo, presumably due to an increase in forest litter (protection from predators) and a decrease in invertebrate predators, although size of adults and clutch size following the hurricane were smaller (Woolbright, 1991). Land snails, in contrast, suffered a severe reduction in population immediately after the hurricane, presumably due to changes in microclimate at the ground level; however, populations rebounded within a few years, as the increased litter augmented food supplies, while maturation of seedlings and saplings provided more shelter (Secrest et al., 1996).

Forest Types and Life Zones

When Columbus arrived on Puerto Rico, it was covered mostly by extensive and luxuriant forests. Almost all contemporary vegetation on the island, however, reflects modifications induced by human activities (Map 3) except a few stands in relictual preserves found within the forest reserves of Luquillo, Carite, Toro Negro, and Maricao. In general, the vegetation of an area is determined primarily by climate, with secondary and important conditioning by physiographic features, such as elevation, drainage, and characteristics of the parent soil (Helmer et al., 2002).

One system for classifying the climax forests of the island—that is, forests that would be present in the absence of human influences—is specific to Puerto Rico. This system recognizes four broad divisions: littoral zone forest, semideciduous subtropical dry forest, tropical and subtropical moist forest, and subtropical rain forest, with each of these containing two subdivisions (Little and Wadsworth, 1964; Picó, 1974; Table 1).

HISTORICAL LAND USE (1969)

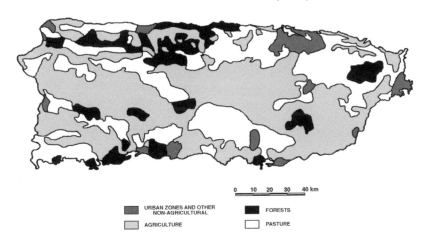

0   10   20   30   40 km

URBAN ZONES AND OTHER
NON-AGRICULTURAL

AGRICULTURE

FORESTS

PASTURE

FOREST RESERVES

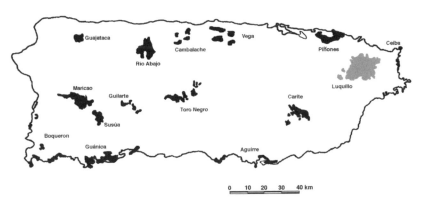

Guajataca
Vega
Cambalache
Rio Abajo
Piñones
Ceiba
Maricao
Guilarte
Luquillo
Toro Negro
Carite
Susúa
Boqueron
Guánica
Aguirre

0   10   20   30   40 km

Map 3 Forests and land use: (a) historical land-use patterns in Puerto Rico in 1969, and (b) location of major forest reserves today (Luquillo = Caribbean National Forest).

Table 1

Common or characteristic species of tree in eight types of forest on Puerto
Rico (Picó, 1974)

Scientific Name	Common Name	
	English	*Spanish*

LITTORAL ZONE FORESTS

MOIST COASTAL FOREST

Acrocomia media	Prickly palm	Corozo
Andira inermis	Angelin	Moca
Calophyllum brasiliense	Maria	María
Hernandia sonora	Jack in the box	Mago
Hymenaea courbaril	West Indian locust	Algarrobo
Manilkara bidentata	Bullet-wood	Ausubo
Mastichodendron foetidissimum	Mastic-bully	Tortugo amarillo
Phoebe elongata	Jamaica nectranda, laurel	Laurel avispillo
Pterocarpus officinalis	Chicken tree	Palo de pollo
Tabebuia heterophylla	White cedar	Roble blanco

MOIST LIMESTONE FOREST

Aiphones acanthophylla	Coyure ruffle-palm	Palma de coyor
Bucida bucerus	Oxhorn bucida, black olive	Úcar
Bursera simaruba	Turpentine tree, gumbo-limbo	Almácigo
Cedrela odorata	Spanish cedar	Cedro hembra
Clusia rosea	Pitch apple	Cupey
Coccoloba diversifolia	Doveplum, pigeon-plum	La uvilla
Coccoloba pubescens	Grandleaf sea-grape	El moralon
Hyeronima clusioides	—	Cedro macho
Licaria salicifelia	Canela	Canelilla
Mastichodendron foetidissimum	Mastic-bully	Tortugo amarillo
Maga grandiflora	—	Maga

SEMIDECIDUOUS SUBTROPICAL DRY FORESTS

DRY COASTAL FOREST

Bucida buceras	Oxhorn bucida, black olive	Úcar
Caparris cynophallophora	Jamaica caper, zebrawood	Burro prieto
Ceiba petandra	Silk cotton	Ceiba
Coccoloba venosa	Chicory grape	Calambrena
Cordia nitida	Red manjack	Capa colorado
Guaiacum officinale	Common lignumvitae	Guayacán
Lonchocarpus domingensis	—	Geno-geno
Pictetia aculeata	Fustic	Tachuelo
Polygala cowellii	Violet tree	Arbol de violeta
Stahlia monosperma	Coabanilla	Cobana negra

DRY LIMESTONE FOREST

Bucida buceras	Oxhorn bucida, black olive	Úcar
Bursera simaruba	Turpentine tree, gumbo-limbo	Almácigo
Crescentia cujete	Calabash tree	Higüero
Guaiacum officinale	Common lignumvitae	Guayacán

Scientific Name	Common Name	
	English	*Spanish*
Guaiacum sanctum	Holywood lignumvitae	Guayacán blanco
Gymnanthes lucida	Oysterwood	Yaití
Pictetia aculeata	Fustic	Tachuelo
Pisonia albida	—	Corcho bobo
Plumeria alba	Milktree	Alelí

TROPICAL AND SUBTROPICAL MOIST FOREST

LOWER CORDILLERA FOREST

Andira inermis	Angelin	Moca
Buchenavia capitata	Wild olive	Granadillo
Cecropia schreberiana	Trumpet tree	Yagrumo hembra
Cordia alliodora	Onion cordia	Capa prieto
Dacryodes excelsa	Candlewood	Tabonuco
Didymopanax morototoni	Matchwood	Yagrumo macho
Guarea trichilioides	American muskwood	Guaraguao
Inga laurina	Sweetpea	Guamá
Linociera domingensis	White rosewood	Hueso blanco
Manilkara bidentata	Bullet-wood	Ausubo
Ocotea leucoxylon	Geo, false avocado	Calcaíllo
Ocotea moschata	—	Nuez moscada
Vitex divaricata	White fiddlewood	Higuerillo

LOWER LUQUILLO FOREST

Cecropia schreberiana	Trumpet tree	Yagrumo hembra
Cyanthea arborea	Tree fern	Helecho gigante
Dacryodes excelsa	Candlewood	Tabonuco
Guarea trichilioides	American muskwood	Guaraguao
Manilkara bidentata	Bullet-wood	Ausubo
Ochroma pyramidale	Balsa	Balsa, guano
Ocotea moschata	—	Nuez moscada
Ormosia krugii	—	Palo de matos
Solanea berteriana	Petit coco	Cacao motillo
Tabebuia heterophylla	White cedar	Roble blanco
Tetragastris balsimifera	Oil tree	Masa

SUBTROPICAL RAIN FOREST

HIGH CORDILLERA FOREST

Alchorina latifolia	Dove-wood	Achiotillo
Calycognium squalulosum	—	Justillo
Clusia krugiana	—	Cupeillo
Cyanthea arborea	Tree fern	Helecho gigante
Cyrilla racemiflora	Swamp cyrilla	Palo colorado
Guatteria blainii	—	Haya minga
Magnolia portoricensis	—	Jaguilla
Micropholis chrysophylloides	Wild star-apple	Caimitillo

Table 1 (continued)

Scientific Name	Common Name	
	English	Spanish

Scientific Name	English	Spanish
Ocotea spathulata	—	Nemoca
Prestoea montana	Sierra palm	Palma de sierra
UPPER LUQUILLO FOREST		
Brunellia comocladifolia	West Indian sumac	Palo bobo
Croton poecilanthus	—	Sabinón
Cyanthea arborea	Tree fern	Helecho gigante
Cyrilla racemiflora	Swamp cyrilla	Palo colorado
Magnolia splendens	—	Laurel sabino
Micropholis garciniaefolia	—	Caimtillo verde
Ocotea spathulata	—	Nemoca
Podocarpus coriaceus	Wild pitchpine	Caobilla
Prestoea montana	Sierra palm	Palma de sierra
Weinmannia pinnata	Wild brazilletto	Oreganillo

Note that some species have no accepted common names in English.

Littoral systems occur along wind-swept sea coasts and include "moist coastal forest" and "moist limestone forest," the latter being found on the well-drained limestone hills of the northwestern karst (Map 4a). Along the south coast, where the rain shadow occurs, semideciduous tropical dry forest exists and comprises extensive areas of "dry coastal forest" (e.g., Fig. 4a, Color Plates) and smaller patches of "dry limestone forest." Tropical and subtropical moist forest is pervasive on lower mountains of the interior; it consists of the "lower cordillera forest," which extends across most of the island, and the "lower Luquillo forest" in the east. Subtropical rainforest on the upper slopes of the mountains includes the extensive "high cordillera forest" and the smaller and wetter "upper Luquillo forest" (e.g., Fig. 4b, Color Plates).

An alternative system of classification divides an area into various "life zones," strictly based on the interaction of temperature and precipitation so that the resulting categories (and types of plants within them) are broadly similar throughout the world (Holdridge, 1947, 1967). Based on the Holdridge system, biologists have described six major life zones for Puerto Rico: subtropical dry forest, subtropical moist forest, subtropical wet forest, subtropical rain forest, lower montane wet forest, and lower montane rain forest (Map 4b; Birdsey and Weaver, 1982; Ewel and Whitmore, 1973).

FOREST TYPES

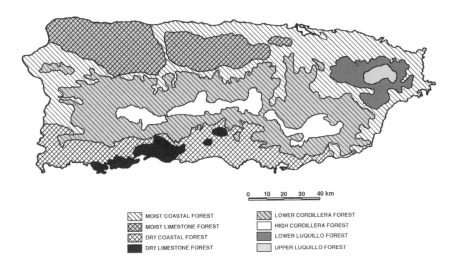

MOIST COASTAL FOREST	LOWER CORDILLERA FOREST
MOIST LIMESTONE FOREST	HIGH CORDILLERA FOREST
DRY COASTAL FOREST	LOWER LUQUILLO FOREST
DRY LIMESTONE FOREST	UPPER LUQUILLO FOREST

LIFE ZONES

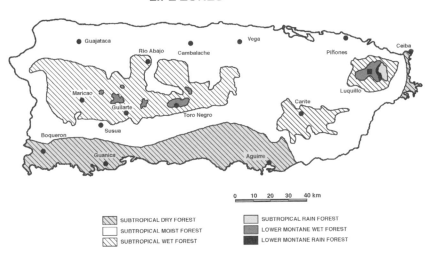

SUBTROPICAL DRY FOREST	SUBTROPICAL RAIN FOREST
SUBTROPICAL MOIST FOREST	LOWER MONTANE WET FOREST
SUBTROPICAL WET FOREST	LOWER MONTANE RAIN FOREST

Map 4 Distribution of (a) forest types (Little and Wadsworth, 1964; Picó, 1974) and (b) life zones in Puerto Rico (Birdsey and Weaver, 1982; Ewel and Whitmore, 1973).

Subtropical dry forest is dominated by deciduous vegetation growing on limestone or saline soils. This zone potentially occupies only 14 percent of the Puerto Rican landscape, occurring along the warm southern coast in an area that receives only 60–110 centimeters of precipitation annually (e.g., Map 4b). In contrast, subtropical dry forest dominates offshore islands, accounting for 68 percent of the potential forest on Vieques and 100 percent of that on Culebra, Desecheo, and Mona. Trees in this zone generally do not exceed 15 meters in height, and crowns are flattened and broad, with sparse foliage. Leaves are small and succulent or tough and leatherlike (coriaceous), and spines and thorns are common. Stands of mangroves that occur in coastal bays or inlets are included in this category.

Subtropical moist forest is the most pervasive life zone on Puerto Rico. Although this forest potentially could cover 61 percent of the main island and 32 percent of Vieques, grasses actually dominate this zone today as a consequence of agriculture (pastures). Annual precipitation in these areas is between 100 and 220 centimeters. Many trees of subtropical moist forest are deciduous; they often attain heights of 20 meters and have rounded crowns. However, areas in southwestern Puerto Rico with serpentine soils harbor a number of endemic species that are characteristically slender, sclerophyllous, and evergreen, with heights typically less than 12 meters. Stands of mangrove also occur in subtropical moist forest, but here the trees attain greater heights than those associated with subtropical dry forest. Remnants of swamp forest, dominated by chicken tree, persist near Dorado, Humacao, La Boquilla, and Vacía Talega, and are part of this life zone.

Subtropical wet forest encompasses 24 percent of Puerto Rico and occurs away from the coast, at higher elevations than the previous zones; consequently, temperatures are somewhat lower, and annual rainfall is greater (200–400 centimeters). This forest has a closed canopy at approximately 20 meters. Species diversity is high (more than 150 species of tree); orchids, epiphytic ferns, and bromeliads are quite common. Serpentine formations in this life zone support more epiphytes, as well as denser and lusher vegetation than are found in subtropical moist forest. When soils are derived from limestone, plants in subtropical wet forest exhibit adaptations more typical of drier conditions, such as small leaves to reduce water loss, thick fleshy parts for water storage, and often spines or thorns for protection against herbivores.

Subtropical rain forest represents only 0.1 percent of the land of Puerto Rico. It occurs on clay soils and is subject to an excess of 380 centimeters of rainfall per year, so soils are saturated continuously. Types of tree in this forest are similar to those of the surrounding subtropical wet forest. Palms, such as sierra palm, and various epiphytes are quite abundant.

Lower montane wet forest is found in cool, high regions, from 700 to above 1,000 meters in elevation, essentially at the summits of most mountains in the eastern and central ranges of Puerto Rico (e.g., Map 4b). It occupies slightly more than 1 percent of the island. Trees are open-crowned, with dark, coriaceous leaves clustered at the ends of branches, giving the canopy a reddish cast. Elfin woodlands—characterized by small, gnarled trees less than 7 meters in height and with evergreen and sclerophyllous leaves—also occur in lower montane wet forest. Stands of palm, especially sierra palm, are found in this life zone as well as in subtropical wet and rain forests.

The final life zone, lower montane rain forest, is the least common, occurring on less than 0.1 percent of the island, along the windward side of the Luquillo Mountains. It supports vegetation similar to that of the lower montane wet forest, except that tree ferns, epiphytes, and palms are more abundant. Elfin woodland occurs on wind-swept slopes and is the most common type of forest in this life zone. Soil surfaces are dominated by superficial and aerial roots, which are laden with bromeliads and liverworts, and the trees are gnarled and festooned with epiphytes.

The Human Influence

The native inhabitants of Puerto Rico (Borikén in their native language) were a group of Arawak Indians called the Taíno. The Arawak were agriculturists and inhabited tropical forests from the foothills of the Andes to the Antilles. The Taíno of Borikén cleared small areas on the island for cultivation of crops, mostly tubers such as cassava (*yuca*), yam, and sweet potato (Jiménez de Wagenheim, 1998). In addition, they harvested *achiote,* corn, *guanábana,* mamey, papaya, pineapple, and *yautía.* Taíno settlements were small and distributed across the island, and thus they caused little disturbance to the landscape, at least relative to what occurred after the European invasion. Around the time of discovery by Spaniards, the Taíno were established in a number of villages, or *yucayeques,* which were scattered throughout the island with major settlements near Cayey, Guayama (or possibly Manatí), Guayanilla Bay, Río Culebrinas, Río Grande de Arecibo, Río Grande de Loíza, Río Naguabo, Río Turabo, Salinas, Toa, and Utuado, as well as on Vieques.

Christopher Columbus first came upon the northwestern shores of Borikén in 1493, perhaps near the site of present-day Aguadilla, and christened the island San Juan Bautista. Colonization of the island by the Spanish, however, did not transpire for another 15 years, when Ponce de León began the conquest in earnest with his landing at San Juan Harbor, which he called Puerto Rico, and founding of a small settlement named Caparra just inland from present-day San Juan (Jiménez de Wagenheim, 1998). Other towns established in the early sixteenth century included Canóvanas, San Germán, and Santiago (Naguabo).

As elsewhere, the Spanish initially sought gold in Puerto Rico, but deposits of the precious mineral were small and were depleted quickly. Consequently the main economic focus became agriculture as early as the 1530s, and it remained the major economic activity until 1955 (Picó, 1974). Sugar was the earliest and most important crop, and it was a mainstay of the economy until well into the twentieth century. At various times ginger and tobacco were also important cash crops, and coffee was introduced to the island in 1736 and became widely planted by 1768. In addition to crops, raising livestock for domestic use and export contributed significantly to the Puerto Rican economy, beginning in the earliest days of the Spanish conquest. Continued expansion of agriculture, however, inevitably led to modifications of the island's ecosystems, unlike anything accomplished by the Taíno.

The pace of deforestation under Spanish rule was slow but continual. Although 850,000 hectares of Puerto Rico were wooded in 1500, the area of forest had decreased 30 percent to 587,000 hectares by 1828 (Table 2). By that time over half the island had been ceded to colonists, primarily for agricultural use, with the other half still protected in Crown forests. Population growth, however, was rapid during the remainder of the nineteenth century, with a concomitant, sharp reduction in forest resources. Almost one million people occupied Puerto Rico at the end of that century, with 90 percent of Crown territory from 1830 having been awarded to settlers and only 182,000 hectares of forest remaining.

Table 2

Temporal trends in area of forest from about 1500 to 1972

| | AREA (1,000 HA) | | |
Year	Forest	Shade Coffee	Total
1500 (ca.)	850	0	850
1828	587	7	594
1899	182	77	259
1912	169	77	246
1916	178	68	246
1931	81	—	>81
1940	—	68	>68
1948	57	57	114
1960	82	—	>82
1972	284	73	357

Compiled by Birdsey and Weaver, 1982

In 1900, after becoming a territory of the United States, only 4 percent of the land was protected in government reserves, and 21 percent was in privately owned forests—most of these located at high elevations and much of their area consisting of "shade coffee." Direct sunlight was harmful to coffee plants, and consequently growers maintained an extensive overstory of mature trees that provided the necessary shade as well as a potential supply of fruit and wood. After 1916, forest area declined rapidly, and by the late 1940s, forests occupied only 6 percent of the island, with shade coffee covering an additional 6 percent. By the mid-twentieth century, Puerto Rico was one of the most severely deforested and eroded regions on earth (Koenig, 1953; Thomlinson et al., 1996).

Recovery of Puerto Rican forests began in the 1950s, as governmental policies favored industrialization, and the economic value of hillside pastures and croplands diminished. Once abandoned, agricultural areas underwent rapid ecological succession and progressed to secondary forests; by 1980, forest represented 30 percent of the island's area (Birdsey and Weaver, 1987). Between 1980 and 1990, secondary forests increased from less than 50 percent to more than 75 percent of the timbered area on the island, partly resulting from a reduction in lands devoted to shade coffee, and by 1992, 42 percent of the island was covered by closed forests (Helmer et al., 2002).

New forests, however, often are not identical in structure and composition to native forests (Lugo and Helmer, 2004). Although use of shade coffee lessened the rate of deforestation, this agricultural practice modified composition of the remaining forest community by selecting for particular species, eliminating some, and introducing others. These patches of modified forest often are the source of seeds for natural regeneration of nearby fields and pastures, and consequently the species composition of the resulting secondary forests often does not mimic composition of native forests. In addition, most new forests occur in small patches, with about 60 percent of regenerating stands occupying less than 0.1 hectare. These highly fragmented forests typically have fewer endemic species, fewer large trees, and less organic matter in the soil than do native forests.

It is not known whether the ongoing recovery of Puerto Rico's forests will prevent extinction of species whose habitats were altered or fragmented in the past, but it is doubtful whether natural food webs ever will be restored completely (Lugo and Helmer, 2004). As a consequence of habitat alteration or destruction by humans, more than 500 species of endemic and non-endemic plants have been classified as rare or endangered or with a restricted distribution. For example, 13 endemic tree species are classified as endangered, and 22 endemic tree species are classified as threatened. Similarly, the wildlife supported by Puerto Rican forests occupies a precarious situation—27 species are considered highly endangered, 29 are endangered, and 15 are on the verge of endangerment (U.S. Department of Agriculture, Soil Conservation Service, and Commonwealth Department of Natural Resources, 1973, 1975; Birdsey and Weaver, 1982).

An Overview of Puerto Rican Mammals

The composition of the fauna of Puerto Rico is a dynamic consequence of three biogeographic and evolutionary processes—immigration or colonization, local extinction, and in situ speciation. As a result of these processes, the number of species (species richness) of mammals on Puerto Rico, like that of most islands, is considerably less than the species richness of comparable areas on the mainland at similar latitudes (Willig and Gannon, 1996). Indeed, the mammalian fauna of Puerto Rico comprises only 13 living, native species, all of which are bats (Table 3; Figs. 5a and 5b, Color Plates). This is far less than the richness of mainland sites of similar size and habitat complexity (mean terrestrial richness of mammals = 83; Table 4).

Table 3

Species of bat currently living on Puerto Rico

Family	Species[1]	Common Name[2]	Approximate Distribution[3]	Conservation Status[4]	Main Item in Diet	Typical Roost
Noctilionidae	*Noctilio leporinus*	Greater bulldog bat	NA, SA, GA, LA	Lower risk, least concern	Fish	Caves
Mormoopidae	*Mormoops blainvillii*	Antillean ghost-faced bat	GA, LA	Lower risk, near threatened	Insects	Caves
	Pteronotus quadridens	Sooty mustached bat	GA	Lower risk, near threatened	Insects	Caves
	Pteronotus parnellii	Parnell's mustached bat	NA, SA, GA	Lower risk, least concern	Insects	Caves
Phyllostomidae	*Brachyphylla cavernarum*	Antillean fruit bat	GA, LA	Lower risk, least concern	Fruit	Caves
	Monophyllus redmani	Greater Antillean long-tongued bat	GA	Lower risk, least concern	Nectar	Caves
	Erophylla sezekorni	Brown flower bat	GA	Lower risk, least concern	Nectar	Caves
	Artibeus jamaicensis	Jamaican fruit bat	NA, SA, GA, LA	Lower risk, least concern	Fruit	Caves
	Stenoderma rufum	Red fig-eating bat	Puerto Rico, Virgin Islands	Vulnerable	Fruit	Foliage
Vespertilionidae	*Eptesicus fuscus*	Big brown bat	NA, SA, GA, LA	Lower risk, least concern	Insects	Caves
	Lasiurus borealis	Red bat	NA, SA, GA	Lower risk, least concern	Insects	Foliage
Molossidae	*Molossus molossus*	Velvety free-tailed bat	NA, SA, GA, LA	Lower risk, least concern	Insects	Buildings
	Tadarida brasiliensis	Brazilian free-tailed bat	NA, SA, GA, LA	Lower risk, near threatened	Insects	Caves

1. Scientific names taken from Koopman (1993) or Simmons (2005).
2. Common names are those that most commonly appear in the literature or are used by workers on Puerto Rico.
3. Species may occupy only part of the indicated areas. NA = North America, SA = South America, GA = Greater Antilles, and LA = Lesser Antilles.
4. Status assigned by International Union for Conservation of Nature and Natural Resources (IUCN; Hutson et al., 2001). Categories used by this organization, listed by decreasing degree of threat to the species, are critically endangered, endangered, vulnerable, and lower risk. Subgroupings within the category of lower risk, in order of perceived threat, are conservation dependent, near threatened, and least concern.

Table 4

Species richness of various taxa for a number of mammalian assemblages in the Neotropics, including Puerto Rico

Taxon	Puerto Rico	Costa Rica							Panama		Colombia			Brazil	
		1	2	3	4	5	6	7	8	9	10	11	12	13	14
Marsupialia	0	6	8	7	5	5	0	6	5	6	—	—	—	4	—
Insectivora	0	0	0	0	2	0	2	2	0	0	—	—	—	0	—
Chiroptera	13	86	81	82	60	45	16	34	29	32	14	16	29	34	25
Emballonuridae	0	9	9	7	4	0	0	0	2	2	0	0	2	1	1
Noctilionidae	1	2	2	2	0	0	0	0	1	1	0	0	0	1	1
Mormoopidae	3	4	4	4	3	1	0	2	1	1	0	0	0	1	1
Phyllostomidae	5	48	44	47	35	34	11	25	22	24	11	12	25	21	14
Natalidae	0	0	0	1	0	0	0	0	0	0	0	0	0	0	1
Furipteridae	0	2	2	0	0	0	0	0	0	0	0	0	0	1	0
Thyropteridae	0	1	1	1	0	1	0	1	0	0	1	0	0	0	0
Vespertilionidae	2	11	9	10	10	8	5	6	2	2	1	2	2	2	4
Molossidae	2	9	10	10	8	1	0	0	1	2	1	2	0	7	3
Primates	0	3	4	3	2	3	0	3	2	3	—	—	—	2	—
Edentata	0	6	7	5	6	3	0	4	5	4	—	—	—	3	—
Lagomorpha	0	1	1	2	1	1	1	1	1	1	—	—	—	1	—
Rodentia	0	17	20	15	16	10	10	16	14	13	—	—	—	10	—
Carnivora	0	13	17	19	12	10	7	14	10	10	—	—	—	3	—
Artiodactyla	0	3	4	3	0	0	1	3	2	2	—	—	—	1	—
Perissodactyla	0	1	1	1	0	0	1	1	0	0	—	—	—	0	—
TOTAL RICHNESS	13	136	143	137	104	77	38	84	68	71	14	16	29	58	25

Sources: Puerto Rico (Willig and Gannon), Costa Rica (Wilson, 1983), Panama (Fleming, 1973), Colombia (Thomas, 1972), and Brazil (Mares et al., 1981, Willig, 1983, and Willig and Mares, 1989).

Site Codes: 1 = La Selva; 2 = Osa; 3 = Guanacaste; 4 = San Jose; 5 = San Vito; 6 = Cerro de la Muerte; 7 = Monteverde; 8 = Balboa; 9 = Cristobal; 10 = Hormiguero; 11 = Pance; 12 = Zabaleta; 13 = Caatinga; and 14 = edaphic Cerrado.

Even if one focuses exclusively on bats, Puerto Rico supports far fewer species than do mainland sites (mean richness of bats = 42; Table 4). The living bats of Puerto Rico shown in Table 3 include one species of fishing bat (Noctilionidae), three mustached or ghost-faced bats (Mormoopidae), five leaf-nosed bats (Phyllostomidae), two plain-nosed bats (Vespertilionidae), and two free-tailed bats (Molossidae). The island does not harbor any species from four New World families—the sheath-tailed or sac-winged bats (Emballonuridae), funnel-eared bats (Natalidae), thumbless bats (Furipteridae), and disk-winged bats (Thyropteridae). Nonetheless, these families contain few species and make only small contributions to the richness of mainland sites. The most notable cause of low species richness of bats on Puerto Rico pertains to the leaf-nosed bats. Even though Puerto Rico harbors five species of phyllostomid, this is less than one-fifth the local richness on the mainland (mean richness of phyllostomids = 27).

Nonflying mammals are not part of the native fauna today, but a number of such animals have been introduced to the main island and its surrounding cays. Native Americans, for example, brought dogs and perhaps guinea pigs to Puerto Rico (Wing, 1989). In addition, Taíno from Hispaniola apparently transported the hutia, a rodent weighing about 1 kilogram, to Puerto Rico as a source of food in pre-Columbian times; remains of this animal are known only from kitchen middens, with no reliable evidence of the animal ever having lived freely on Puerto Rico (Miller, 1918; Olson, 1982).

European settlers also intentionally imported a number of domesticated mammals: dogs, cats, goats, pigs, cattle, horses, and donkeys (Lawrence, 1977; Miller, 1929; Wing, 1989; Wing et al., 1968). Some introduced species exist in a feral state, even to this day, either because escaped animals have adapted to local conditions and populations are self-perpetuating or because domesticated individuals that stray or are abandoned continually enter feral groups, thereby rescuing marginally fit populations from extinction. The potential ecological impact of introduced animals is well known and quite detrimental (e.g., overgrazing by goats; Breckon, 2000), especially on smaller islands and cays. In other situations, such as heavily forested sites on Puerto Rico, feral cats and dogs have an unknown but possibly appreciable impact on indigenous animals that did not evolve in the presence of any member of the cat or dog families (Garcia et al., 2001; Hawkins, 1998; Willig and Gannon, 1996).

Unlike domesticated species, some mammals were introduced inadvertently by humans. The black rat likely arrived in Puerto Rico as a stowaway with Ponce de León in 1508 (Snyder et al., 1987), and it ultimately was joined on the island by the house mouse and Norway rat. These mammals flourished wherever food, water, and shelter were available, and today they are found throughout Puerto Rico—in urban, rural, and natural areas. Worldwide, these commensal rodents are responsible for large economic losses due to destruction and contamination of human food and are a potential health hazard (Jackson, 1982). In addition, introduced rodents, particularly rats, may significantly alter community composition and structure of food webs in natural settings, like the Luquillo Experimental Forest, because of their large body size and potentially high densities (Willig and Gannon, 1996).

Although domesticated species occasionally escape captivity and rats and mice hitchhike their way to new lands, the Javan mongoose actually was released intentionally by humans into the wild on Puerto Rico, in 1877, to prey on rats living in sugarcane plantations (Horst et al., 2001). Mongooses, however, are not capable of regulating rat populations, and once established, these carnivores generally alter the ecological structure of natural habitats that they invade. For example, six species of vertebrates are considered endangered on Puerto Rico, partly because of interactions with rats and mongooses (Raffaele et al., 1973). Negatively affected species include the Puerto Rican boa, Puerto Rican vine snake, Puerto Rican parrot, Puerto Rican short-eared owl, and possibly the Key West quail-dove and Puerto Rican whip-poor-will.

Although the only native mammals living on Puerto Rico today are bats, there is fossil evidence for the past presence of a number of other species, including bats and nonflying mammals. These extinct animals include one shrew, one sloth, three leaf-nosed bats, and five rodents (Table 5). Cause of each extinction remains poorly understood, but as on other islands of the Antilles, disappearance of a few species may be related to climatic changes at the end of the Pleistocene Epoch (Pregill and Olson, 1981). Many recent extinctions, however, apparently were induced by humans, either by alteration of habitats and exploitation by natives and Europeans or by interactions with rats, cats, dogs, and mongooses (Morgan and Woods 1986; Woods and Eisenberg, 1989).

Table 5
Extinct mammals of Puerto Rico

Order	Family	Common Name of Family	Species
Insectivora	Nesophontidae	West Indian shrews	*Nesophontes edithae*[1]
Xenarthra	Megalonychidae	West Indian sloths	*Acratocnus odontrigonus*[1, 2]
Chiroptera	Phyllostomidae	Leaf-nosed bats	*Macrotus waterhousii*[3]
	Phyllostomidae	Leaf-nosed bats	*Monophyllus plethodon*[4]
	Phyllostomidae	Leaf-nosed bats	*Phyllonycteris major*[1]
Rodentia	Capromyidae	Hutias	*Isolobodon portoricensis*[1, 5]
	Echimyidae	Spiny rats	*Heteropsomys insulans*[1]
	Echimyidae	Spiny rats	*Homopsomys* [= *Heteropsomys*] *antillensis*[1]
	Echimyidae	Spiny rats	*Puertoricomys* [= *Proechimys*] *corozalus*[1]
	Heptaxodontidae	Giant hutias	*Elasmodontomys obliquus*[1, 6]

Sources: Anthony 1918, 1926; Baker and Genoways, 1978; Best and Castro, 1981; Choate and Birney, 1968; Morgan and Woods, 1986; Nowak, 1999; Ray, 1964; Whidden and Asher, 2001; White and MacPhee, 2001; and Woods, 1989.
1. Extinct from all known areas; island endemic.
2. Includes *Acratocnus major* (Hall, 1981).
3. Extinct on Puerto Rico, although currently living on other islands and the mainland.
4. Extinct on Puerto Rico, although currently living on other islands; island endemic.
5. Native to Hispaniola and may have existed only in captivity on Puerto Rico.
6. Includes *Heptaxodon bidens* (Ray, 1964).

Island Biology

Beyond the well-documented pattern that the number of species on islands is reduced compared with equivalent areas on the mainland, one of the most pervasive biogeographic patterns is a reduction in species richness with reduction in area of an island. Indeed, the "equilibrial theory of island biogeography" (MacArthur and Wilson, 1967) formalizes the view that number of species on an island represents a balance between rates of loss (primarily extinction) and rates of gain (primarily colonization and in situ speciation). Size of an island affects extinction rates because area should be proportional to the number of animals that the island can support and to the number of habitats that it contains. All other things being equal, small islands should have small populations, and species with small populations should have a greater likelihood of going extinct, thereby diminishing the richness of a particular island.

Colonists must come from somewhere, and the distance from such "source pools" (e.g., the mainland) should be inversely proportional to the rate at which individuals, and hence new species, colonize an island. Therefore, all other things being equal, islands that are close to source pools should have more species than do islands that are distant from source pools. In addition, islands close to source pools should have fewer endemic taxa, which usually result from in situ speciation, because immigrants from adjacent populations are constantly arriving, thus preventing the genetic isolation needed for speciation to occur.

The Caribbean region contains a plethora of islands of various sizes and distances from the mainland, providing a rich source of variation in the primary factors thought to mold species richness. At the same time, different islands have different geological histories, elevational profiles, primary source pools (North America versus Central or South America), secondary source pools (nearby islands that might provide colonists), and exposure to anthropogenic and natural disturbances. Such uncontrolled differences among islands lead to considerable complexity, sometimes confounding the ability to detect patterns related to size or distance. Nonetheless, at broad levels of resolution, the way in which bats, birds, amphibians, and reptiles differ in richness among the islands of the Greater Antilles is generally consistent (Fig. 6). Specifically, bat and amphibian faunas are always more depauperate than avian and reptilian faunas, with species richness of bats being the lowest of the four groups on all islands.

The number of species of bat on islands of the Antilles is reasonably consistent with predictions based on size of the island (Griffiths and Klingener, 1988). Indeed, 87 percent of the variation in bat species richness in the Antilles is explained by variation in island area alone (Fig. 7). The species richness of bats on Puerto Rico, however, is slightly below the predicted value for an island of its size in the Antilles, as is that of Hispaniola, whereas richness for Cuba and Jamaica are above predicted values. These differences among islands likely are related to lower sea levels during the Pleistocene Epoch. As a result, Florida and the Grand Bahama Bank were larger, providing North American bats with easier access to Cuba. Similarly, the Nicaraguan Plateau and the Yucatan Peninsula were significantly larger, providing shorter over-water routes for Central American species colonizing Jamaica, and to a lesser extent, Cuba. Puerto Rico and Hispaniola, in

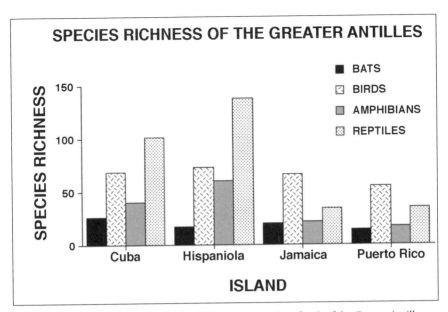

Fig. 6 The abundance of bats, birds, amphibians, and reptiles of each of the Greater Antilles (M. R. Willig and J. Alvarez, unpublished data).

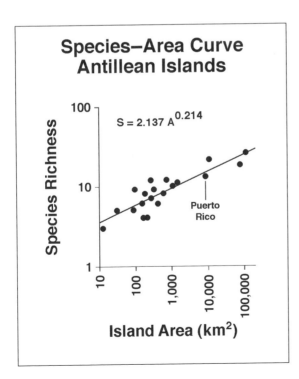

Fig. 7
Relationship between bat species richness (S) and area in kilometers^2 (A) for islands in the Greater and Lesser Antilles (Griffiths and Klingener, 1988).

contrast, were relatively isolated from both sources of colonists, even during the Pleistocene Epoch.

Bats probably colonized Puerto Rico via "stepping-stone dispersal"—that is, movement first from the mainland to a nearshore island, then to a more distant island, and so on (Griffiths and Klingener, 1988). The ancestors of most species initially left Central or North America and occupied the westernmost of the Greater Antilles, with some eventually flying farther east to Hispaniola and Puerto Rico. Although colonization of Puerto Rico could have been accomplished by animals traveling from South America northward through the Lesser Antilles, biogeographers generally consider this route less likely for two reasons (Koopman, 1989). First, for those species of bat that currently live both on the mainland and on the Antilles, individuals from the Greater Antilles generally are more similar morphologically to those of Central or North America than they are to their relatives in South America. Second, in those genera that are endemic to the West Indies, there is higher species richness in the Greater than in the Lesser Antilles, suggesting a longer period of residence in the Greater Antilles.

Community Ecology within the Caribbean Basin

Whatever the initial origin, exactly which species successfully colonized and ultimately survived or evolved on each island in the Caribbean was not due totally to chance, because the resulting communities of bats are not random samples of the species available from mainland source pools (Rodríguez-Durán and Kunz, 2001). Instead, biogeographers have identified a core community of bats, the members of which are similar on most islands of the West Indies. This core community consists of one species of Antillean long-tongued bat (genus *Monophyllus*), one species of Antillean fruit bat (*Brachyphylla*), the Jamaican fruit bat, greater bulldog bat, Brazilian free-tailed bat, and velvety free-tailed bat. On the larger islands that are part of the Greater Antilles, the core community expands to include three species of ghost-faced or mustached bat (*Mormoops* and *Pteronotus,* respectively) and one species of flower bat (*Erophylla*). Puerto Rico contributes to this general pattern, with 10 of the 13 species of bat on the island representing members of the core community (Table 3).

Why did this particular combination of species evolve on many different islands in the West Indies? One factor that probably played a role in determining the makeup of these island communities was partitioning of food resources (Rodríguez-Durán and Kunz, 2001). Although a random sample of species from the mainland theoretically could include only frugivores or only insectivores, for example, the core community that exists on Caribbean islands typically contains species that specialize on different foods, including nectar/pollen (long-tongued and flower bats), fruit (Antillean and Jamaican fruit bats), fish (greater bulldog bat), and insects (Brazilian free-tailed, velvety free-tailed, ghost-faced, and mustached bats).

Another key factor that helped shape community composition on West Indian islands was type of roosting site utilized by each species (Rodríguez-Durán and Kunz, 2001). Except for the velvety free-tailed bat, all members of the core community roost either predominantly or exclusively in caves. In addition, over 40 percent of the 56 living species of bat that occur in the West Indies (80 percent of species on Puerto Rico) use caves. These underground roosts provide excellent protection from wind and rain, especially during tropical storms, such as hurricanes. More than 100 hurricanes have ravaged at least some part of the region over the last 500 years (Colón, 1977), and species that roost in more exposed situations, such as tree hollows or foliage, likely suffer greater direct mortality from these common storms than do cave-dwelling species (e.g., Gannon and Willig, 1994).

In addition, islands usually have fewer food resources than do comparable sites on the mainland (e.g., Janzen, 1973). Consequently, reptiles, birds, and mammals that live on islands often are characterized by having lower rates of energy expenditure (mass-specific basal metabolic rates) than do their mainland relatives. These low metabolic rates presumably are preadaptations that favored some island-colonizing groups over others, or these low rates of energy expenditure are adaptations evolved by members of endemic taxa in response to the low level of resources available on islands (Faaborg, 1977; Kurta and Ferkin, 1991; McNab, 1994, 2001).

Many cave-dwelling bats of the West Indies, especially endemic species (e.g., flower bats, ghost-faced bats, Antillean long-tongued bats), prefer to roost in "hot caves," where temperatures often exceed 30°C (Silva-Taboada, 1979). Because these species spend most of their lives in very warm environments, heat loss is minimal, and they typically possess a reduced rate of heat production—that is, a low basal metabolic rate (Bonaccorso et al., 1992; Rodríguez-Durán, 1995). This physiological trait presumably allows such species to attain higher population densities on islands, making extinction less likely despite the low availability of resources. Similarly, insectivorous species of bat have lower mass-specific basal rates of metabolism than bats having other dietary preferences (McNab, 1982), and this characteristic may partly explain why insectivorous species are more common in the core community of bats on West Indian islands than are species that primarily consume fruit, nectar, or fish.

Bats are the only mammals capable of true flight. Other so-called flying mammals, such as flying squirrels (order Rodentia) and flying lemurs (order Dermoptera), are excellent gliders, but they are not able to achieve sustained, powered flight. The ability to fly is an important element in the evolutionary success of bats, and today bats comprise the second most diverse order of mammals, the Chiroptera. There are about 1,100 species of bat worldwide, which means that one out of every five mammalian species is a bat (Wilson and Reeder, 2005). Moreover, bats make a disproportionately large contribution to the high diversity of species in tropical areas (e.g., Kaufman and Willig, 1998; Willig and Selcer, 1989; Willig et al., 2003), and bats often are the most abundant group of mammals within ecological communities (Patterson et al., 2003).

The order Chiroptera generally is divided into two suborders—the Megachiroptera and the Microchiroptera (but see Teeling et al., 2002). The Megachiroptera contains only one family, the Pteropodidae or Old World fruit bats, which are restricted largely to the tropics of Asia and Africa. The Microchiroptera, conversely, has essentially a worldwide distribution, occurring on all continents except Antarctica and on many oceanic islands. The suborder Microchiroptera also is more diverse than the Megachiroptera and contains 17 families and about 90 percent of all bat species, including all those found on Puerto Rico.

Some biologists believe that the two suborders of bats evolved independently; that is, the order Chiroptera is not monophyletic. This view implies that mammalian flight evolved twice and that the broad similarities between megabats and microbats, such as structure of the wing, are the result of convergent evolution. Proponents of this concept suggest that megachiropterans actually are related more closely to the order Primates, whereas the nearest relative of bats was historically believed to be the order Insectivora. Although some evidence supports the contention that flight evolved twice in mammals, it is not the prevailing view (Jones et al., 2001).

In any event, people have long mistakenly believed that bats are related to rodents, particularly mice, and this misconception is readily seen in words that are used for "bat" in different languages—*fledermaus* or "flying mouse" in German, *chauve-souris* or "bald mouse" in French, and *murciélago* or "blind mouse" in Spanish (actually derived from Latin and old Castilian Spanish and not the modern word for mouse). However, the controversy surrounding the origin of bats is based on a disagreement over whether all bats evolved from insectivores or whether the Megachiroptera descended from primates. Insectivores and primates are more closely related to each other than they are to rodents, so no matter which explanation is correct, bats actually may be more closely related to humans in an evolutionary sense than they are to mice.

Structure of a Bat

Bats are small mammals, with a median body mass of approximately 20 grams (Jones and Purvis, 1997). The smallest bat in the world is the hognosed bat of Thailand; it weighs only as much as a single penny, about 1.5–2 grams, and it is arguably the smallest mammal in the world. Even the largest bat is quite small compared with the giants of the mammalian world, such as elephants, bears, and whales. The largest bat is the Malayan flying fox, which weighs up to 1,500 grams, or about as much as a typical rabbit. Despite its low mass, a Malayan flying fox has a wingspan up to 1.5 meters (Kunz and Jones, 2000).

Although the anatomy of bats appears strange, bats are constructed using the basic body plan of all mammals (Fig. 8). The wing of a bat, for example, has the same skeletal components that are found in a mouse or shrew or human. The major difference is that in a bat the bones of the arm and hand are greatly elongated to support the wing membrane, which is

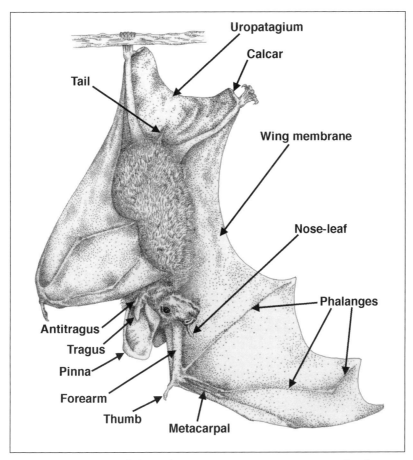

Fig. 8 Stylized drawing of a bat, indicating important structures mentioned in the text. Drawing by K. Rizzo.

just a supple piece of skin that extends from the side of the body. The name of the order, Chiroptera, is Greek for "hand-wing" and refers to the fact that the distal part of the wing membrane is supported by four fingers and the associated metacarpals. The first finger, the thumb, is much shorter than the others and is tipped by a sharp claw. The thumb moves independently of the wing membrane and is used primarily as a "hook" that helps the bat climb a tree or the wall of a cave.

Bones of the hind limbs also are elongated compared with those of other mammals, although less so than bones of the forelimb (Fig. 8). Unlike a songbird, which flies with its legs held close to the belly, a bat flies with its long legs extended backward to support a tail membrane (uropatag-

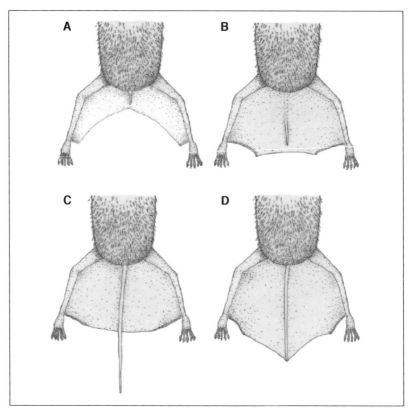

Fig. 9 Types of tail membranes typical of (a) leaf-nosed bats (phyllostomids), (b) ghost-faced or mustached bats (mormoopids), (c) free-tailed bats (molossids), and (d) plain-nosed bats (vespertilionids).

ium). The tail membrane is another piece of skin that stretches between the legs and rump and often encloses some or all tail vertebrae. A tail membrane is developed best in insectivorous species and is least extensive in frugivorous and nectarivorous bats (Fig. 9). The uropatagium provides extra aerodynamic lift and helps with aerial maneuvers. In addition, a well-developed tail membrane often is used as a basket to catch the young at birth (e.g., Kurta and Stewart, 1991) or to capture flying prey (Griffin et al., 1960).

Although the legs of bats are long and thin, their feet and toes are similar to those of other mammals. Each of the five toes of a bat ends in a curved claw, and the bat uses these for grooming and to anchor itself while hanging upside down in its roost. Bats hang for many consecutive hours each day, but this behavior requires little energy (Schutt, 1993). Tendons that control the grasping action of the digits can be locked anatomically into place in a manner analogous to that used by perching birds (order Passeriformes). It actually requires more energy for the animal to release its grip than to continue hanging.

Why do bats hang upside down? One hypothesis is that the hind legs lengthened over evolutionary time to provide support for the tail membrane, but simultaneously, the bones became quite thin to minimize the load that the flying animal must carry. Ultimately, the bones became too thin (weak) to bear the weight of the animal's body in a typical standing position, making it advantageous to adopt a less stressful hanging habit (Howell and Pylka, 1977).

Seeing with Sound

Virtually all bats are nocturnal, with most activity occurring between approximately sunset and sunrise each day. Some bats, particularly the Old World fruit bats, have large eyes and rely totally on sight to navigate and avoid obstacles in the dark. Microchiropterans typically have smaller eyes, but contrary to popular belief, no bat is blind. The eyes of bats are designed to function under conditions of low light, and bats can see better than humans in dimly lighted environments (Altringham and Fenton, 2003). Although microchiropterans have eyes, they rely on sound, not light, to guide them through the darkness.

Microchiropteran bats use echolocation, a form of sonar (not radar), in much the same way that a ship's crew would search for a submarine or sunken vessel. A pulse of sound is emitted in the direction of the bat's movement, and the animal listens for the returning echoes, just as ship personnel would send a pulse toward the bottom of the sea and track the returning signal. Because the speed of sound in air or water is more or less constant, a bat's brain or ship's computer can calculate the distance to the object. By emitting a train of successive pulses and analyzing the resulting echoes, one can assess the speed and direction of a moving target, whether it is a flying insect or an attacking submarine. A bat, to a certain degree, also can distinguish among objects based on differences in the frequency or amplitude of the incoming echo. Such differences may result, for example, from distinctive patterns of wing beats used by different potential prey.

Frequency of the echolocation pulse usually is between 20 and 120 kilohertz; however, some bats use frequencies as high as 200 kilohertz, and for others, they may be as low as 10 kilohertz. The upper limit of human hearing is only 20 kilohertz. Most echolocation pulses therefore are inaudible to us, and we often are not aware of a bat flying by in the dark. Even though people cannot hear these pulses, they may be quite intense (loud), depending on species (Fenton, 1982). Pulses emitted by the big brown bat, for example, are louder than a typical smoke detector (105 decibels at 10 centimeters).

Why use such high frequencies? The frequency of any sound is inversely proportional to its wavelength, so high frequencies mean short wavelengths. The best sound for detecting an object is one that has a wavelength about the same size as the object itself. Because many bats feed on small insects and all bats must detect small items like twigs in their environment, the best sounds are those that have short wavelengths and, unavoidably, high frequencies. For example, the wavelength of a sound with a frequency of 100 kilohertz is only 3.4 millimeters, making it suitable for detecting even tiny insects. Although high-frequency sounds allow a bat to detect small things, the energy contained in such sounds is absorbed rapidly by the atmosphere and by objects in the environment. Consequently, bats are able to detect small items using echolocation only at rather short distances, often within only a few meters.

Echolocation pulses are classified into three broad types based on the relationship of frequency and time, although there is much variation (Schnitzler and Kalko, 1998). Some species of bat typically emit calls of constant frequency, whereas many others use frequency-modulated calls— that is, pulses that change or "sweep" through a range of frequencies from beginning to end. In a third type of call, both frequency-modulated and constant-frequency components occur (Fig. 10).

The physics of sound and the anatomy and physiology of bats interact in such a way that constant-frequency (narrowband) calls are well suited for simple detection of a target, but they are not ideal for localizing an object in space; the reverse, however, is true for frequency-modulated (broadband) calls (Schnitzler and Kalko, 1998). Type of signal emitted and specific frequencies that are used by a bat are influenced by many factors, including body size, foraging habitat, foraging strategy, and type of food (Bogdanowicz et al., 1999; Fenton, 1982; Jones and Rydell, 2003).

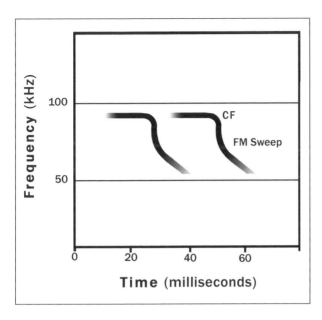

Fig. 10
A stylized graph of an echolocation call (frequency versus time), illustrating the concepts of constant frequency (CF) and frequency modulation (FM).

In the last 15 years, inexpensive "bat detectors" have become a popular research tool for biologists, greatly expanding the ability to study echolocation and foraging behavior of bats in the dark. These handheld electronic instruments have a microphone that is sensitive to the ultrasonic frequencies used by bats. Once the microphone detects a high-frequency echolocation pulse, the instrument emits a sound of lower frequency that humans can hear. Thus the detector allows one to know when bats are flying by even when they cannot be seen, and in some circumstances, one can identify the bat to species based on the timing of the pulses and the frequencies that the bat uses (Barclay, 1999; O'Farrell et al., 1999; Parsons et al., 2000).

Nocturnal Behavior

Bats probably evolved from nocturnal ancestors, but why do they continue to restrict their activity to the nighttime? To a large degree, continued nocturnal behavior of bats may be related to risk of predation from hawks and falcons (Speakman, 1995). Bats are relatively slow-flying animals and are easy prey for such swift, agile birds. In addition, most bats rely on echolocation, which is not very useful for detecting a fast-moving predator at a distance.

Additional factors that may cause bats to avoid diurnal activity are the risk of overheating (Thomson et al., 1998) and perhaps dehydration. Bat wings are naked, highly vascularized, and have a large surface area. A flying bat, like any exercising mammal, generates a substantial amount of heat. When flying at night, a bat floods its wings with blood so that they act as radiators, losing heat to the cooler surroundings and nighttime sky, thus maintaining a normal body temperature. During the day, however, those same wings would absorb solar radiation, especially in tropical environments, and the animal would actually gain heat through the wings and run the risk of fatal hyperthermia. Consequently, continued nocturnal activity may be a means of avoiding this potentially deadly problem of overheating.

Shelter and Insulation

Most bats spend at least 12 hours each day at rest, and during this time, they must find protection from inclement weather as well as a place to hide from potential predators. In addition, young bats cannot forage until they are able to fly, and most cannot fly until they reach about 70 percent of adult size (Barclay, 1994). During this prolonged period of development (Barclay and Harder, 2003; Jones and MacLarnon, 2001), youngsters must remain in a secure place for 24 hours each day. Consequently, availability of suitable shelter is crucial to survival of any species of bat and an important determinant of species diversity in a given geographic area (Humphrey, 1975).

Shelters occupied by bats are quite variable (Kunz and Lumsden, 2003). Although people may be most familiar with bats that roost in buildings, few species ever use houses, churches, or barns. Instead, bats most commonly roost in caves, rock crevices, and trees. Trees provide a wealth of roosting opportunities; many bats hide in cavities or underneath exfoliating bark, and other species, often cryptically colored, cling to the trunk or hang among the leaves. Some species, such as the long-eared myotis of western North America, do not even need an entire tree but will roost in stumps left behind after logging (Vonhoff and Barclay, 1997).

Most bat species simply take advantage of secluded sites that already exist and do not modify their environment to make a suitable shelter. One species, the New Zealand lesser short-tailed bat, occasionally may excavate burrows within fallen trees (Daniel, 1979). Perhaps 20 species of bat, from both the New World and Old World, construct temporary "tents" from leaves, stems, or root masses of tropical plants (Kuntz and Lumsden, 2003; Kunz et al., 1994; Kunz and Lumsden, 2003). A leaf tent, for example, results when these bats use their teeth to sever parts of a living leaf until it collapses around the animal and encloses the bat on all sides and from above. Various plants are acceptable for tent-making; the most commonly used species are palms, philodendrons, bananas, and heliconias.

Although most small mammals, such as mice and shrews, construct a nest from dry grass, shredded bark, fur, or other insulating material, no bat builds a nest within its roost. Nevertheless, many roosting sites, including tree hollows or solution cavities in the ceiling of a cave (Fig. 11, Color Plates), trap warm air that rises from the bat's body. Because a bat generally hangs from the top of its roost enclosure, it receives thermal benefits by resting where the warm air accumulates. Many species of bat roost in small groups, with individuals often in contact with one another; this behavior provides additional energetic benefits by minimizing heat loss when the animals cluster and by increasing the amount of warm air that accumulates (Kurta, 1985). It is possible for an entire cave to act as a heat trap, if its entrance is small and low relative to interior chambers. If large numbers of bats occupy such a cave, accumulated heat may result in ambient temperatures of 35°C or more year-round.

If bats do not build nests, then why do they fly into human hair? The answer, of course, is that bats have no need of or attraction for hair. It is possible that at some time in history, a bat became entangled in the hair of a human, but if so, it was an accident for which the bat probably paid with its life. There are many examples of bats being caught by barbed wire or cactus spines that punctured the wing membranes, trapping the bats and ultimately leading to their death (Barbour and Davis, 1969). Surely these bats were not seeking the barbed wired or the cactus spines to build a nest either. Most likely, the bat simply flew too close to a barb or spine and became accidentally trapped, and the same probably is true for the bats-and-hair story. A bat was either chasing an insect past someone's head or perhaps trying to escape from a human in a narrow attic or small cave. As the bat flew by, the claw on its toe or thumb inadvertently snagged a wayward hair, the bat was caught, and another myth about bats began.

Food, Guano, Ecology, and the Economy

Worldwide, diets of bats are quite diverse. Some species feed on fish, land crabs, frogs, lizards, rodents, other bats, leaves, and even the blood of birds and mammals. Most bats, however, concentrate on one of three foods: insects, fruit, or nectar and pollen. Fruit- and nectar-feeding bats are primarily restricted to the tropics, but insect-eating bats are found throughout tropical and temperate areas. About two-thirds of all bat species consume insects, whereas most others subsist on plant products. Although the remarkable diversity and adaptations of bats suggest their ecological importance, many people do not realize that bats provide a number of critical services to ecosystems simply by eating and processing their food. In particular, bats are agents of pollination, seed dispersal, and insect control, and these roles make bats "keystone taxa," especially in subtropical and tropical areas.

For example, seedlings of many tropical trees and vines, such as figs, are unable to grow successfully in the shade of their parents (Fleming, 1982; Heithaus, 1982). Frugivorous bats, however, consume the ripe fruit of these plants, and by the time undigested seeds are defecated, the bat is often many meters or even many kilometers away from the parent tree (Fig. 12, Color Plates). Passage of seeds through the digestive system of bats enhances the rate of germination for some species of plant (Fleming, 1988). In tropical areas like Puerto Rico, bats are particularly important in providing the seeds necessary to revegetate areas of forest destroyed by windstorms or landslides, such as those caused by hurricanes, and in bringing native seeds to agricultural fields and pastures that have been abandoned. Clearly, bats and their diurnal counterparts, birds, play pivotal roles in the recovery of biotic communities after a natural or human-induced disturbance (Willig and McGinley, 1999).

Many tropical plants also are pollinated by bats, including century plant, wild banana, and silk cotton. Some of these plants have coevolved with bats so that the pattern of flowering and morphology of the flower are discouraging to diurnal pollinators yet favorable to bats (von Helversen and Winter, 2003). Bat-pollinated flowers, for example, often have a strong, musky odor that is attractive to flying mammals, and the flowers are not colored brightly, thus minimizing their attraction to visually oriented birds and insects. Such flowers generally are open only at night, when bats are active, and the flowers often are large and located around the periphery of the tree, making access easier and safer in the dark. Some bat-pollinated flowers even have "acoustic mirrors"—structures that strongly reflect ultrasonic sounds produced by a bat, making it more likely that the animal will discover the nectar-filled flower (von Helversen and von Helversen, 1999). On Puerto Rico, about half the species of bat are primarily frugivorous or nectarivorous, and they depend on a variety of local and introduced plants for their food (Appendix 1).

The role of bats in controlling insect populations is equally important yet often underappreciated. For example, a female bat during late lactation often consumes her own weight in insects each night (Kurta et al., 1989, 1990), and a single colony of bats on Puerto Rico may consume over 20 tons of insects every month (Rodríguez-Durán and Lewis, 1987). Not only do bats limit the number of insects in general, but these flying predators often feed heavily on pests that damage crops and forests. Big brown bats, for instance, help control adults of the corn rootworm (Whitaker, 1995), and Brazilian free-tailed bats consume huge numbers of corn earworm moths every night (McCracken, 1996). As we learn further details about the diet of different types of bats, the role of insectivorous species in protecting plants that are economically important to humans surely will become more evident.

Bats also are critical for the functioning of unique and little-known ecosystems that exist underground, in caves, such as those in the northwestern karst region of Puerto Rico and on Mona Island (Culver, 1982; Wilkens et al., 2000). In the absence of light and primary producers (plants), cave ecosystems require the input of energy- and nutrient-bearing material from outside the cave, and this is a service provided by bats (Willig and McGinley, 1999). Cave-dwelling bats often forage large distances from their roosts and, upon returning at dawn, shower the floor of their cave with a rain of feces and urine.

Waste materials from live bats, as well as the bodies of dead bats that fall to the floor of the cave, provide the raw material of life for a plethora of small-to-microscopic organisms that dominate these complex lightless ecosystems (Nieves-Rivera, 2003; Peck, 1974, 1981). These subterranean communities contain a diversity of feeding guilds, including fungivores and detritivores and their specialized predators (e.g., Conn and Marshall, 1991; Trajano and Gnaspini-Netto, 1994; Whitaker et al., 1991). In some parts of the world, such cave environments are among the most stressful to life because of high and sometimes toxic levels of nitrogen-containing compounds (e.g., ammonia) that can accumulate there (Studier et al., 1967).

In addition to its ecological function, bat guano has played a role in the economic development of several parts of the world, including Puerto Rico. Populations of cave-dwelling bats can number in the millions, and significant amounts of guano accumulate over long periods. This nutrient-rich material (Studier et al., 1994) has long been used as an organic fertilizer by people throughout the world (Hutchinson, 1950). On Mona Island, for example, about 150,000 metric tons of guano were mined for fertilizer between 1877 and 1927 (Frank, 1998). Exhaustion of deposits and increased reliance on cheaper synthetic fertilizers spelled an end to guano mining on Mona Island and in other parts of the United States, but bat guano still is mined and utilized extensively in underdeveloped countries, especially in Southeast Asia (Hutson et al., 2001).

Bats and Pre-Columbian Residents of Puerto Rico

A number of history books suggest that in pre-Columbian times, the early inhabitants of Puerto Rico used bats as a source of food (Rodríguez-Durán, 2002). This theory is based primarily on a comment attributable to Gonzalo Fernández de Oviedo, who wrote in 1535 that the Taíno boiled bats to remove hairs from the flesh and that the resulting meat was white in color and delicious (Fernández Méndez, 1981). Fernández de Oviedo, however, never actually witnessed the preparation and consumption of bats, and his statement is countered by a number of facts. For example, the large flight muscles of bats are red, not white. In addition, remains of bats have never been found in native middens on any island of the Caribbean, except Saba in the Netherlands Antilles (Wing, 2001), and no other culture in the New World utilized bats for food. Thus it seems unlikely that bats were ever a substantial component of the diet of the Taíno or earlier inhabitants of Puerto Rico.

Nevertheless, bats were important to the early human settlers of Puerto Rico, as shown by frequent incorporation of these flying mammals into the pottery and artistic carvings of the time (Rodríguez-Durán, 2002). Although some representations can be identified only broadly as leaf-nosed bats, others can be recognized as brown flower bats, Antillean fruit bats, or velvety free-tailed bats (Fig. 13, Color Plates). The existence of numerous and detailed renderings of bats in stone, conch, and pottery suggests that these animals were held in high esteem by native peoples, and it now seems more likely that bats played a spiritual rather than a dietary role in the lives of early inhabitants of Puerto Rico.

Vampire Bats and the Vampire Myth

Vampire bats really do exist, and these mammals are some of the most fascinating results of evolution (Greenhall and Schmidt, 1988). There are three species: common, hairy-legged, and white-winged vampire bats. All are moderately sized, generally weighing between 30 and 45 grams, and they live only in the New World, from northern Mexico to south-central South America. Although fossilized specimens are known from Cuba, no vampire bat currently occurs on any of the Antilles. Their historic absence from most of the Antilles, including Puerto Rico, may be related to the paucity of moderate-to-large mammals and birds that could serve as prey on the islands.

Each species of vampire subsists only on blood. Common vampire bats prefer mammalian blood, with cattle the most frequent prey. In contrast, the other two species feed heavily on birds, especially poultry, but they may attack mammals as well. Vampire bats do not suck blood; rather, they nick the skin with their highly modified and razor-sharp incisors and lap the oozing blood with their long tongue. To prevent clotting, saliva of vampire bats contains an anticoagulant that allows blood to continue flowing for several minutes or more while the bat feeds.

Vampires also have a thermal (infrared) sense that can distinguish differences in temperature between parts of the prey's body without having to touch and possibly alert the intended victim. Warmer areas of skin have a better flow of blood, and apparently the bats use this infrared sense to help decide where to bite. Bites do occur occasionally in the neck area, but they also occur on the toes, feet, nostrils, and even around the margin of a bird's cloaca.

The myth of a human vampire may be as old as civilization, and it can be traced at least as far back as the Babylonian "Edimmu," a blood-sucking soul (Murphy, 1991). The actual word *vampire* apparently comes from Slavic folklore, which depicts vampires as troubled souls leaving their dead bodies at night, wandering the countryside, and sucking blood from sleeping humans (Villa-C. and Canela-R., 1988). The stories initially had the undead assuming the form of dogs, cats, horses, and even birds, and the association with bats did not occur until long after the expeditions of Cortez revealed the existence of blood-feeding bats in the New World.

Bats and Public Health

Worldwide, several human diseases are linked to wild animals (Feldhamer et al., 2004). For example, fleas associated with ground squirrels in the western United States are still a source of the bacteria that result in bubonic plague, and throughout much of North America, ticks that infest other terrestrial mammals spread bacteria that cause Rocky Mountain spotted fever or Lyme disease. Hemorrhagic fever, in contrast, does not require a parasitic intermediary; this human disease is induced by a hantavirus obtained by contacting the saliva or urine of small rodents, such as the deer mouse. Although no known disease of humans is transmitted by the ectoparasites of bats (Constantine, 1988), bats are implicated in the spread of two potentially serious illnesses: histoplasmosis and rabies.

Histoplasmosis

Histoplasmosis is a disease usually associated with the lungs, although other tissues may be affected (Constantine, 1988). It is caused by a fungus, *Histoplasma capsulatum,* which can exist in two forms, either as a mycelial saprophyte living on rich organic matter or as a yeast parasitizing internal tissues of various mammals. *Histoplasma capsulatum* occurs naturally in moist soil, but it also thrives in bird or bat guano under humid conditions. The fungus is present in most tropical and warm-temperate regions of both the Old World and New World, including Puerto Rico and 31 of the 48 contiguous United States.

Histoplasma capsulatum reproduces through spores that are deposited in the soil or guano, and humans most often contract the disease when dried guano is disturbed and spores are inhaled. After inhalation, the spores transform into the parasitic yeast stage. Most infections are mild and go unnoticed, and about 20 percent of adult Puerto Ricans unknowingly have experienced these asymptomatic infections (Beck et al., 1976; Carvajal Zamora, 1977). However, about 10 percent of those infected develop flu-like symptoms, such as difficulty breathing, unproductive cough, or chest pain. Fever, loss of weight, and night sweats also may occur. Severity of the disease, which can be fatal, depends on number of spores that are inhaled. The disease is obtained only by breathing in spores from the soil or guano and cannot be contracted by direct contact with an infected human, bat, or other mammal.

On Puerto Rico, severe infections can be obtained by visiting certain bat caves, such as the Aguas Buenas Caves (Beck et al., 1976), where large amounts of guano exist and ventilation is very poor. Anyone entering such caves should wear a tightly fitting respirator with a filter capable of eliminating particles as small as 2 microns.

The simple presence of bats in a building, on the other hand, should not make one fear histoplasmosis. The fungus does not survive well in hot, dry roosts, such as attics, and disturbing guano in these locations is not likely to result in an infection (Tuttle and Kern, 1981). Nevertheless, anyone attempting to remove accumulated guano from a building or otherwise spending a prolonged time in poorly ventilated areas containing bird or bat guano should also wear an appropriate respirator as a precaution.

Rabies

Rabies is a fatal disease of the central nervous system that is caused by a bullet-shaped virus in the family Rhabdoviridae (Brass, 1994; Messenger et al., 2003). The virus proliferates in the salivary glands of an infected animal and most often is transmitted to an uninfected individual through a bite. Once an animal is infected, the virus spreads along the sheath surrounding peripheral nerves, eventually entering every organ. After an incubation of several weeks or more, infected individuals display changes in behavior, such as loss of appetite, irritability, and increased sensitivity to sound or light. Eventually the disease progresses to either a furious or a paralytic phase.

In the furious form, the victim becomes aggressive, wandering about and biting any animal that it encounters, whereas in the paralytic phase, the victim slowly loses control over its muscles and dies without any unusual signs of aggression. The phase that occurs depends to a certain degree on the type of animal that is infected. Dogs, for example, frequently experience the furious phase, whereas bats most commonly experience the paralytic version of the disease. Although some animals may survive initial exposure to the virus, individuals that shed the virus in their saliva or develop other symptoms ultimately die; no animal can act as an asymptomatic "carrier," transmitting the virus to others yet remaining free of the disease itself.

Rabies and related viruses occur worldwide (Brass, 1994; Messenger et al., 2003; Rupprecht et al., 2002). In developing countries, dogs are the main reservoir of the disease, but in more affluent regions like the United States and Canada, dogs are vaccinated routinely. Consequently, in these countries, wild mammals—foxes, raccoons, skunks, and bats—are the principal remaining reservoir of the virus (e.g., Krebs et al., 2001).

Rabies has been detected in over 30 species of bat in the United States alone and in at least 30 additional species living elsewhere in the New World (Constantine, 1988). In the United States and Canada, about 3–10 percent of bats turned in to health departments for testing are positive for the rabies virus. This does not mean that 3–10 percent of all bats are rabid, because sick bats are more likely to be caught and tested than are healthy bats. Surveys based on bats captured while leaving their roosts indicate a positive rate of less than 0.5 percent for most species (Constantine, 1988). Despite the widespread nature of rabies in bats, human fatalities from bat rabies are exceedingly rare, with less than 40 cases in the United States during the last 50 years (Rupprecht et al., 2002); one has a greater chance of dying from a lightning strike or lawnmower accident than by contracting rabies from a bat.

Although widespread on the mainland, rabid bats are very rare on the Antilles and are reported from only Grenada and Cuba (Price and Everard, 1977; Silva Taboada and Herrada Libre, 1974). On Puerto Rico, introduced mongooses, dogs, and cats are the animals that most commonly are infected with rabies. For example, between 1977 and 1992, 939 rabid mammals were discovered on Puerto Rico; 76 percent of these were mongooses, 13 percent were dogs, and 5 percent were cats (R. L. González-Peña, personal communication).

To date, no rabid bat has been detected on the island (R. L. González-Peña, personal communication), but given the small number of bats on

Puerto Rico that have been tested and presence of the disease in bats on Cuba, we suggest that humans avoid handling bats whenever possible, especially bats that appear helpless or paralyzed. If bitten by a bat or any other mammal, one should save the animal for testing by the Department of Health and immediately seek medical advice, because the disease can be prevented easily with prompt administration of a rabies vaccine. Animal control workers, biological researchers, or others whose activities unavoidably bring them into frequent contact with bats should consider a pre-exposure vaccination as a precaution (Bernard et al., 1987).

Eliminating Bats from Buildings

Many people are understandably reluctant to share their home or work place with a colony of bats. Accumulated guano may cause an objectionable odor, and bats are active and noisy at night when people sleep. Moreover, the species that live in buildings are responsible for most bat-human bites, simply because these animals are more likely to come into contact with people than are bats that roost far away in trees or caves. Although mothballs, ultrasonic noisemakers, and poisonous chemicals have been advocated as means of removing bats from buildings, these techniques rarely if ever work. Potential conflicts between bats and humans are resolved best by "bat-proofing" the building—that is, eliminating the small holes that bats use to enter the house (Greenhall, 1982; Williams-Whitmer and Brittingham, 1995).

On Puerto Rico, only one species, the velvety free-tailed bat, typically roosts in houses. Because this bat leaves its roost before it is completely dark, a homeowner can determine how bats are entering and exiting the house by watching the roofline for about one hour, starting just before sunset. These openings can be covered temporarily at dusk, after all bats have left and before they return from feeding, using plastic sheeting or hardware cloth. During the light of day, the homeowner can accomplish a more permanent closure using standard masonry or carpentry techniques.

Bats do not gnaw like a mouse and are not likely to reopen a properly sealed entrance. They may, however, have a number of alternative entrances that they begin using after the preferred opening is closed. Consequently, a homeowner may need to watch the house and cover holes for a number of nights before all openings are eliminated. In any event, bats should not be evicted from a building when they are reproductively active; doing so decreases the reproductive success of pregnant bats (Brigham and Fenton, 1986) and leads to death of flightless young that are left behind when their mothers are prevented from returning.

Family Noctilionidae
Bulldog or Fishing Bats

The family Noctilionidae contains a single genus with only two species—the greater bulldog bat and the lesser bulldog bat (Simmons, 2005). Although the family is small, these medium-to-large bats live in much of the Neotropics, ranging from central Mexico to northern Argentina as well as on many islands in the West Indies and Bahamas. Throughout their range, bulldog bats are found most often in coastal areas and other lowlands, and these bats always live near sources of water, such as streams, lakes, or oceans. Bulldog bats generally gather in groups of less than 100 individuals and roost in caves, rock fissures, hollow trees, and occasionally buildings.

Noctilionids are quite distinctive. Their upper lip is separated at the midline by a vertical fold of tissue, giving them a "hare-lip" appearance. In addition, they have a well-developed nose-pad, a chin with noticeable cross ridges, and very large canines. These combined features give noctilionids an appearance that vaguely resembles a bulldog, and this similarity is the source of one of the common names for the family.

The alternative common name refers to an important item in the diet of the greater bulldog bat: fish. Even though there are about 1,100 different species of bat, fish eating regularly occurs only in a few species representing three different families—Vespertilionidae, Megadermatidae, and Noctilionidae (Aihartza et al., 2003; Ma et al., 2003; Stadelmann et al., 2004). The greater bulldog bat gaffs fish with its huge hind feet, in a manner somewhat similar to that used by a bald eagle or osprey. The lesser bulldog bat, in contrast, rarely eats fish, although it may use its hind feet to pluck insects from the surface of a pond or stream. Indeed, some biologists hypothesize that grabbing insects from the water's surface was an evolutionary first step in the development of catching fish (Kalko et al. 1998; Lewis-Oritt et al., 2001; Siemers et al., 2001). The bulldog bat is the only fish-eating species that has invaded the Antilles and lives on Puerto Rico.

Noctilio leporinus (Linnaeus, 1758)
Greater Bulldog Bat or Greater Fishing Bat
Murciélago Pescador

Taxonomy: *Noctilio leporinus* was first described by Linnaeus almost 250 years ago using a specimen captured in Surinam. Today taxonomists recognize three subspecies: *N. l. leporinus, N. l. mastivus,* and *N. l. rufescens. N. l. mastivus* is the subspecies found on Puerto Rico, and its original description was based on a bat taken on St. Croix in the Virgin Islands.

Name: The word *noctilio,* at least partly, is based on the Latin word for night or nocturnal (Jaeger, 1955). The specific epithet *leporinus* also comes from Latin and combines two words that mean "like a hare," probably referring to the somewhat elongate ears and split upper lip (hare-lip) of this bat. *Mastivus,* in contrast, is a latinized English word for "hound" and is a reference to the bulldoglike appearance of this animal.

Linnaeus was the first to note the harelike appearance of this species and constructed the word *leporinus* (Hood and Jones, 1984). Vahl, 40 years later, described a seemingly new species of fishing bat from St. Croix and used the term *mastivus* as the specific epithet to denote the bat's resemblance to a dog. Ultimately, taxonomists realized that both animals were the same species; hence the older term *leporinus* was retained as the specific epithet, but *mastivus* was used as the subspecific name for fishing bats living on the Antilles. That is why today there is a bat on Puerto Rico that is called the "doglike harelike animal of the night."

Distribution and Status: Overall, the range of the greater bulldog bat extends from Sinaloa and Veracruz, Mexico, through Central America

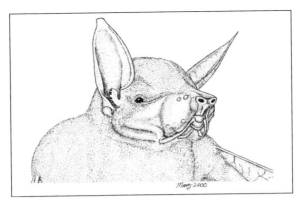

Fig. 14 Portrait of a greater bulldog bat.

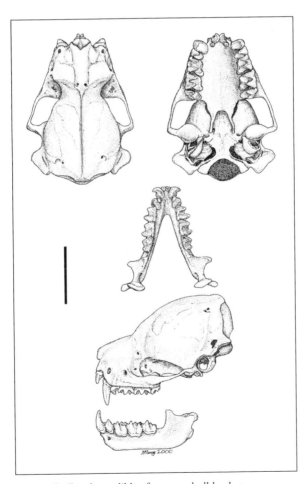

Fig. 15 Skull and mandible of a greater bulldog bat
(bar = 10 mm).

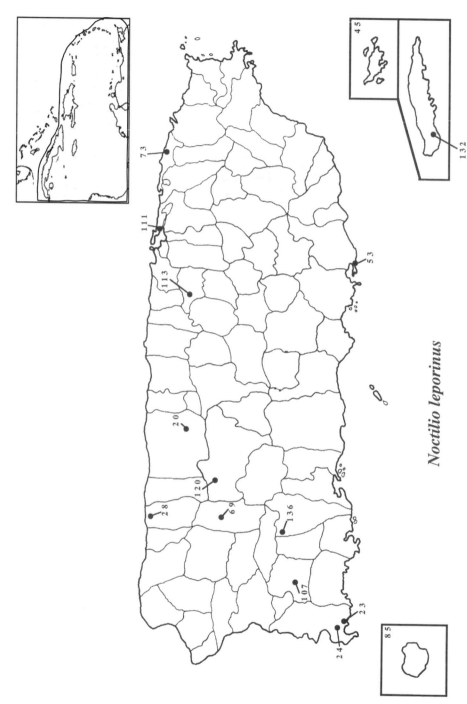

Map 5 Distribution of the greater bulldog bat on Puerto Rico and in the Caribbean basin. Location of numbered sites is detailed in Appendix 8.

Noctilio leporinus

and into South America, as far south as northern Argentina (Hood and Jones, 1984; Map 5). The species, however, is not distributed evenly throughout this broad area. It is generally absent from mountainous zones and those that are extremely arid and is most likely to be found along the coast, in lowlands, and along major river systems, including the Amazon. The greater bulldog bat also inhabits all major islands of the Caribbean Sea from Trinidad to the Bahamas to Cuba. The subspecies *N. l. mastivus* occupies these islands, as well as the mainland from Mexico to northern South America, whereas the other subspecies occur farther south on that continent.

The greater bulldog bat was documented first on Puerto Rico in 1916, on the outskirts of Loíza, east of San Juan (Anthony, 1918). Since that time few specimens have been collected, and they are from only a dozen-or-so localities on Puerto Rico and the islands of Mona and Vieques. This bat most often is seen flying along the coast, where it forages for fish at dusk, but we also have observed bulldog bats feeding over large inland bodies of water, such as Lago Guajataca and Lago dos Bocas in the western part of Puerto Rico, and even along small streams near Maricao. It is likely that systematic sampling over inland rivers and ponds and areas of quiet water along the coastline would yield additional records.

Measurements and Dental Formula: Total length is 112–130 mm; length of tail, 23–33 mm; length of hind foot, 26–34 mm; height of ear, 22–30 mm; and length of forearm, 85–90 mm. Body mass is 48–77 g. Animals from Culebra, however, appear smaller than those from the main island. For example, body mass for 10 males and 10 females from Camuy is 77.4 ± 6.2 and 65.2 ± 4.4 grams (*SD*), respectively, whereas body mass for 10 males and 10 females from Culebra is 54.8 ± 6.9 and 47 ± 4.3 grams, respectively (A. Brooke, personal communication). Dental formula is incisors 2/1, canines 1/1, premolars 1/2, molars 3/3 = 28.

Description: The greater bulldog bat is among the largest of New World bats and the largest living on Puerto Rico. The split upper lip, large nose-pad, ridged chin, and large canines make this bat unmistakable (Figs. 14, 15). Ears are long, narrow, and pointed, and the outer margin of the tragus is serrated. The hind limbs, which are used to capture prey, are obviously elongate and equipped with very long toes and claws. The tail is shorter than the interfemoral membrane, and although most of the tail is wrapped in the uropatagium, the last few millimeters protrude from the top of the membrane. Dorsal fur is short and ranges in color from pale yellowish orange to dark orange or from orange-brown to grayish brown. A white middorsal stripe extends from head to rump. As in most bats, ventral hairs are paler, and in the greater bulldog bat, the belly varies in color from dirty white to orange.

The skull is quite large, with greatest length over 26 millimeters. The only other species on Puerto Rico with skulls that are so large are the Jamaican and Antillean fruit bats, in the family Phyllostomidae. Phyllostomids, however, have two distinct openings in the palate behind the incisors (incisive foramina), but such openings do not occur in noctilionids (Fig. 15). Hence the combination of large size and lack of incisive foramina distinguishes the skull of the greater bulldog bat from that of all other species on Puerto Rico.

Natural History: Greater bulldog bats prefer to spend the day in caves, rock crevices, or hollow trees, but they sometimes roost in manmade structures (Brooke, 1994, 1997; Hood and Jones, 1984). Other species occasionally use the same roost as the greater bulldog bat. For instance, mustached bats, ghost-faced bats, Greater Antillean long-tongued bats, and Antillean fruit bats may share sea caves with this species on Cuba (Silva-Taboada, 1979), and Jamaican fruit bats and Greater Antillean long-tongued bats will roost with the bulldog bat on Puerto Rico (Rodríguez-Durán, 1998). The different species, however, generally keep to themselves and seldom interact.

Colonies containing up to several hundred greater bulldog bats are known, but most groups on Puerto Rico consist of 30 or fewer individuals (Brooke, 1994, 1997; Hood and Jones, 1984; Nowak, 1999). Within the roost, bulldog bats cluster in small groups containing two to nine adult females and a single adult male (Brooke, 1997). Bachelor males roost alone or with one to four other males. The male that is found with the females typically is older and has larger testes than do the bachelor males, and this resident male seems responsible for most copulation. Individual males may monopolize a cluster of females for two or more years before being replaced by a bachelor male. These "harems" are quite stable in composition. Although females occasionally visit other groups within the dayroost, they generally return to their original cluster, and each multifemale group occupies the same site within a roost over many years.

Both sexes produce an oily yellow fluid that accumulates on the skin and fur under the wings (Brooke and Decker, 1996). The secretions are quite odiferous, and it is possible for some humans to identify a bulldog bat on the basis of odor alone as it flies by in the dark. While roosting, females from the same cluster apparently coat themselves and each other, but not members of another group, with these secretions. Chemical content of secretions varies between sexes and individuals, and these scents may communicate information concerning individual identity, sexual identity, and/or reproductive condition.

Bulldog bats are active throughout the night, with individual bouts of foraging lasting 0.75–1.5 hours and total foraging time adding to 2.5 hours per night (Brooke, 1994, 1997). Some females hunt in groups, exiting the dayroost individually but eventually flying off with other females; resident males do not accompany females. Individuals may return to the same foraging area on consecutive nights and use the same site in successive years. Between foraging bouts, bulldog bats rest and digest, either at the dayroost or at a site distant from where they spend the day. Bats from a number of different dayroosts may intermingle for short periods in these nightroosts, although each returns to its traditional dayroost before dawn.

The greater bulldog bat is known best for its habit of eating fish, and like all bats in the New World, it relies on echolocation, even when hunting this unusual prey. Ultrasonic pulses consist of a short constant-frequency element at about 54 kilohertz, followed by a long frequency-modulated sweep down to about 28 kilohertz (O'Farrell and Miller, 1999; Schnitzler et al., 1994). Such high-frequency sounds are either absorbed by the water or reflected from its surface and actually are useless for detecting a fish underwater. Nonetheless, surface-dwelling fish frequently cause ripples, occasionally break the surface of the water with their bodies, and sometimes jump into the air while escaping from underwater predators. It is only then that the fish is detected by an echolocating bat patrolling just above the water.

If a fish is detected, the bat rapidly strikes by snapping its hind feet down and forward into the water, unlike an eagle, which strikes down and back. At other times, the bat simply trolls for fish, dragging its feet through the water for up to 10 meters in areas with high fish activity; when the bat trolls, the bottom of the feet face forward and instantly grab any fish that is contacted (Altenbach, 1989; Bloedel, 1955; Schnitzler et al., 1994). Captured fish are passed quickly to the mouth. There the prey is chewed and swallowed or stored temporarily in internal cheek pouches while the bat flies to a nightroost for more leisurely processing of its meal (Murray and Strickler, 1975). Not surprisingly, roosting sites typically have a fishy odor that emanates from accumulated feces and discarded fish parts, although some of this odor also may be attributed to the bats' skin secretions (Brooke and Decker, 1996). A single bat consumes as many as 30 or 40 fish, each 25–76 millimeters long, in a single night. On Culebra, diet of the greater bulldog bat includes eight species of fish, each representing a different family (Brooke, 1994). Tilapia is the most common fish prey on that island, followed by silversides and ballyhoo.

Although best known for eating fish, the greater bulldog bat actually consumes a variety of animal foods (Hood and Jones, 1984). On Culebra this species preys heavily on insects and to a lesser extent on shrimp, terrestrial crabs, and scorpions (Brooke, 1994). Beetles and moths are the most common insect prey, with crickets, flies, true bugs, and roaches also consumed. Bulldog bats forage for insects over ponds and fields, along roads, and at streetlights. Flying insects are scooped out of the air using the wing or tail membrane, but terrestrial prey, in contrast, are grabbed with the hind feet in a manner similar to when the bats are catching fish. Thus these bats have the behavioral flexibility to capture prey from the water, ground, or air. Fish and insects are consumed throughout the year; however, insects are the most important dietary item in the wet season (July–December), whereas fish are more prevalent in the dry season (January–June).

Reproductive data for this bat on Puerto Rico are sparse, but an excellent data set exists for Cuba. On that island, gestation occurs from December to April, with pregnant females in April containing a near-term fetus (Silva-Taboada, 1979). Although pregnant females are not found in May or June, a few with very small embryos have been found in July and August. This suggests that most bulldog bats on Cuba give birth only once each year, but a small proportion of the population may breed twice in the same year. The second, smaller peak in reproduction does not occur in all parts of this bat's range (e.g., Brazil; Willig, 1983) and may not happen on Puerto Rico. On Culebra, Brooke (1997) notes that copulation occurs from September to November, but young are not observed until May. She speculates that the nine-month interval between mating and parturition indicates some kind of reproductive delay, such as delayed fertilization or delayed implantation.

Litter size is only one. The naked neonates weigh about 12–13 grams each or 21 percent of the mother's mass (Silva-Taboada, 1979). Pups never hang alone during the day, although mothers leave their offspring behind while the adults forage (Brooke, 1997). Young bulldog bats do not take their first flight until they are almost two months old. Captive individuals may live up to 11.5 years (Wilkinson and South, 2002).

The only parasite known from this animal on Puerto Rico is a streblid fly, *Paradyschria fusca* (Anthony, 1918). A large number of ectoparasites, in contrast, have been reported from other areas, including labidocarpid, sarcoptid, macronyssid, and spinturnicid mites, argasid ticks, and batbugs, as well as streblid flies (Hood and Jones, 1984). Endoparasites include a few flukes and roundworms.

Family Mormoopidae
Mustached or Ghost-faced Bats

Mormoopids live in a variety of habitats ranging from desert scrub to tropical forests and are found from southern Arizona and Texas in the United States to Mato Grosso in Brazil. Although the Mormoopidae and the Phyllostomidae differ in many characters, recent molecular evidence and analysis of internal and external anatomical traits indicate a close relationship between these families (Simmons and Conway, 2001; Van Den Bussche et al., 2002). The family Mormoopidae is small, with only eight extant species and two genera. Five of the eight species are widespread in the Antilles, and three are present on Puerto Rico.

Mormoopid bats are small or medium-sized bats (forearm length 35–65 millimeters) that are quite distinctive. Lips of a mormoopid are thickened, and depending on species, there are a number of stiff hairs, bumps, flaps, and other excrescences surrounding the mouth and nostrils. External ears of a mormoopid curve ventrally and pass beneath small, inconspicuous eyes. The tail membrane is well developed, much better so than in a leaf-nosed (phyllostomid) bat, and the membrane typically is about twice as long as the tail itself. Moreover, the terminal portion of the tail protrudes from the dorsal side of the membrane when a mormoopid is at rest, but the vertebrae disappear into the uropatagium when the legs are spread, as in flight (Fig. 9).

Hairs of mormoopids are short and densely packed over the body. The pelage ranges from dark brown or grayish black to orange, even within a single species. In general, intraspecific variation in color of bats may be controlled genetically, may be related to molting patterns, or may result from a bleaching effect caused by accumulation of noxious gases, particularly ammonia, in enclosed roosts. However, noticeably high levels of ammonia, such as those found in some roosts of free-tailed bats (Studier et al., 1967), are uncharacteristic of mormoopid roosts on Puerto Rico.

During the day, mormoopids typically roost in caves, forming large colonies containing thousands and often hundreds of thousands of individuals from a number of different species. Inside the cave, different species usually are segregated, such that each taxon occupies a different chamber or different section of a chamber (Silva-Taboada, 1979). Caves favored by the Mormoopidae are the so-called "hot caves," in which air temperatures typically range from 28 to 40°C. Most deep caves in Puerto Rico have a temperature of only 21°C, and higher temperatures in hot caves result from a combination of a large number of bats, decomposing guano, and a small entrance that is lower than the rest of the cave. Inside the cave, each bat and the microorganisms living in the guano act as living furnaces that unavoidably produce a small amount of heat as a by-product of chemical reactions continually occurring within their bodies. This internally produced heat is transferred to surrounding air molecules; the warmed air then rises and becomes trapped in the cave because the small size and low position of the entrance inhibit exchange of warm internal air with the cooler outside environment.

Moormopids that roost in hot caves are prone to rapid dehydration when outside their roosts, despite a kidney structure that favors water retention (Rivera-Marchand and Rodríguez-Durán, 2001). However, hot caves are moist as well as warm places, with relative humidity typically exceeding 90 percent. High moisture levels are caused by dripping water within the cave, but continual evaporation of water from the animals' bodies, feces, and excreted urine also supplies moisture to the stagnant air. These species likely congregate in large groups because of the energetic benefits derived from roosting in such a stable, warm environment, as well as the benefits of reduced water loss during the long day-roosting period (Rodríguez-Durán, 1995).

Mormoopids feed exclusively on insects that they capture in flight. These bats have long, narrow wings that are very lightweight, especially compared with wings of more generalized bats, such as phyllostomids or noctilionids (Vaughan and Bateman, 1970). The reduced mass of a mormoopid wing may be an adaptation that effectively increases speed of flight and improves maneuverability, allowing these bats to commute long distances to foraging grounds and to pursue and capture flying insects.

Mormoops blainvillii Leach, 1821
Antillean Ghost-faced Bat
Murciélago Barbícacho

Taxonomy: The genus *Mormoops* contains only two living species—the island-dwelling *M. blainvillii* and the larger *M. megalophylla,* which is primarily restricted to the mainland from Texas to South America. *M. blainvillii* is monotypic; that is, there are currently no recognized subspecies (Simmons and Conway, 2001). The original description of *M. blainvillii* is based on a specimen captured in Jamaica.

Name: The generic name *Mormoops* comes from the Greek words *mormo* and *opis.* A rough translation of these words is "having the appearance of a hideous monster," and they allude to the unusual facial features of these bats. The specific epithet *blainvillii* honors a French physician and naturalist of the early nineteenth century, Henri M. D. de Blainville (Lancaster and Kalko, 1996).

Occasionally the scientific name *Aello cuvieri* is incorrectly used for the Antillean ghost-faced bat (e.g., Hall, 1981). This occurs because the species was described and named twice in the same publication, with *A. cuvieri* appearing eight pages earlier in a different article than *M. blainvillii* (Lancaster and Kalko, 1996; Leach, 1821). The original descriptions of these two "species" mistakenly indicated differences in dental formula and other features, but more than 50 years after the original publication appeared, Dobson (1878) examined the holotype of each and concluded that they were identical. Although the name *A.*

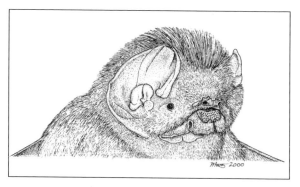

Fig. 16 Portrait of an Antillean ghost-faced bat.

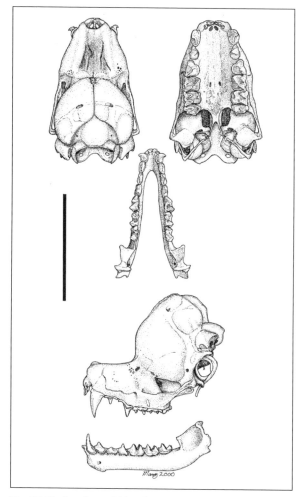

Fig. 17 Skull and mandible of an Antillean ghost-faced bat
(bar = 10 mm).

cuvieri technically appeared first in the literature, the accompanying description of the species was incorrect, and consequently the name typically was not used by biologists. The description of *M. blainvillii,* however, was correct, and that name had by then been applied many times to the Antillean ghost-faced bat. For this and other reasons, the International Commission on Zoological Nomenclature officially recognized the name *M. blainvillii* and suppressed use of *A. cuvieri* (Rezsutek and Cameron, 1993). Although Simmons (2005) indicated that the correct spelling of the specific epithet was *blainvillei,* this spelling has not appeared in any scientific literature pertaining to the Antillean ghost-faced bat, and we chose to maintain the commonly used *blainvillii* to avoid confusion.

Distribution and Status: Presence of fossilized specimens indicates that at one time the range of this species extended from the Bahamas as far south as Antigua and Barbuda in the Lesser Antilles (Lancaster and Kalko, 1996). Today, however, this mormoopid is known only from the Greater Antilles—Cuba, Jamaica, Hispaniola, and Puerto Rico (Map 6). The ghost-faced bat is found throughout Puerto Rico and on nearby Mona Island and is very abundant.

The Antillean ghost-faced bat is listed as a species of "low risk" by the International Union for Conservation of Nature and Natural Resources (IUCN), although it is also considered to be "near threatened" (Hutson et al., 2001). This latter designation indicates that there is a high risk of extinction in the medium-term future and that there is no management program aimed at conservation of this ghost-faced bat. The species is considered near threatened not because of a low number of individuals but because it is found only on a small number of islands and because these bats roost in large colonies in very specialized sites— hot caves. Destruction of a single one of these unusual roosts could result in the death of tens or even hundreds of thousands of individuals.

Measurements and Dental Formula: Total length is 78–87 mm; length of tail, 21–31 mm; length of hind foot, 6–10 mm; height of ear, 11–18 mm; and length of forearm, 46–50 mm. Body mass is 8–11 g. The dental formula is identical in all mormoopids: incisors 2/2, canines 1/1, premolars 2/3, molars 3/3 = 34.

Description: The ghost-faced bat is a rather small bat, especially compared with some phyllostomids, such as the Jamaican fruit bat or Antillean

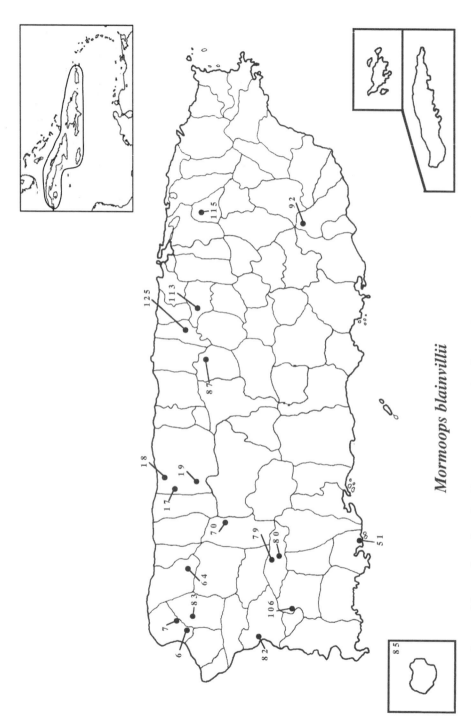

Map 6 Distribution of the Antillean ghost-faced bat on Puerto Rico and in the Caribbean basin. Location of numbered sites is detailed in Appendix 8.

Mormoops blainvillii

fruit bat, and "it is easily the most grotesque in appearance of any of the bats on Puerto Rico" (Anthony, 1918:347). The snout of a ghost-faced bat is short and, as in all mormoopids, lacks a nose-leaf (Fig. 16). Nasal openings and upper lips are combined into a complex labionasal plate, with an irregular outline and fleshy outgrowths lateral to the nasal openings. The lower lip consists of a single fleshy flap with a bumpy (papillose) surface that is separated from a pair of lower flaps. Coarse bristles are associated with both upper and lower lips.

External ears are short but very broad. An inconspicuous band connects the ears across the top of the head, and the lower edge of the external ear is continuous with the lower lips, thus giving the impression of a conical funnel leading to the ear opening on either side of the face. The tragus is large and complex compared with those of phyllostomids and vespertilionids. Dorsal fur is long, soft, and dense and varies in color from pale or amber brown to bright orange; ventral hairs are shorter and slightly paler. Wings and tail membrane are dark and essentially naked. As in all mormoopids, the tail is shorter than the uropatagium, and the terminal end of the tail protrudes from the top of the tail membrane when the bat is at rest.

With a greatest length of 13–15 millimeters, the skull of the Antillean ghost-faced bat is smaller than that of any other species on Puerto Rico except the sooty mustached bat. Ghost-faced bats, however, possess incisive foramina (Fig. 17), which are lacking in mustached bats. In addition, the skull of the ghost-faced bat is unique among Puerto Rican bats in that the rostrum forms an angle of nearly 90° with the forehead.

Natural History: The Antillean ghost-faced bat roosts mostly in hot caves, where each bat typically hangs by itself from the ceiling. Individuals occasionally gather in small clusters in cooler sections of caves, and sometimes solitary bats are found in an apparent state of shallow torpor in places where air temperature may fall as low as 20°C (Rodríguez-Durán, 1995). In hot caves, ghost-faced bats attain high densities. For instance, an estimated 50,000 of these bats roost in Cucaracha Cave near Aguadilla (Rodríguez-Durán and Lewis, 1987).

This mormoopid normally begins to exit the cave after dark, between 22 and 55 minutes after sunset (Silva-Taboada, 1979). Although some ghost-faced bats begin leaving at this time, the exodus does not peak until two to three hours later, and bats may enter and leave the cave throughout the night (Rodríguez-Durán and Lewis, 1987; Silva-Taboada, 1979). In caves of mixed-species composition, ghost-faced bats typically begin departing after sooty mustached bats but at about the same time as Antillean long-tongued bats. When flying, ghost-faced bats often produce a soft, eerie, quavering sound that is somewhat reminiscent of distant spaceships in a science fiction movie; the sound presumably results from vibrations of the membranes or other body parts as the bat moves through the air.

The Antillean ghost-faced bat is a swift-flying aerial hunter that, like other insectivorous bats, uses its large tail membrane and wings to pluck insects from the air. Moths are by far the most common food, and on some nights, so many of these insects are captured that flight membranes of the bats become coated with the microscopic scales of their prey. Ninety percent of 65 fecal samples from Puerto Rico contained moths (excluding pellets with no identifiable remains; Rodríguez-Durán and Lewis, 1987), and 100 percent of 49 stomachs from Cuba held these insects (Silva-Taboada, 1979). In addition to moths, beetles occur in about 28 percent of fecal pellets examined, whereas true bugs, flies, and leafhoppers are found in less than 20 percent (Rodríguez-Durán et al., 1993). Moths typically are more active late in the night rather than near sunset, and ghost-faced bats emerge from their roost later than most other bat species; the combination of these behavioral patterns likely explains the dominance of moths in the diet (Jones and Rydell, 1994; Rodríguez-Durán, 1984).

Foraging behavior has not been studied in detail. These bats appear to hunt along forest edges and around the crowns of trees (Schnitzler et al., 1991; Silva-Taboada, 1979), and we assume that individuals from large colonies commute long distances to their foraging grounds each night to minimize competition.

An animal's diet plays a role in how well it is able survive a natural disaster, such as a hurricane. Insectivorous species like the Antillean ghost-faced bat are less affected by these storms than are frugivorous or nectarivorous species, presumably because insects are less susceptible to destruction than are flowers and fruits and the trees that bear them

(Barlow et al., 2000; Jones et al., 2001). For example, ghost-faced bats represented less than 10 percent of the 750,000 bats living in Cucaracha Cave before Hurricane Georges struck the island in 1998, but this proportion increased to over 75 percent in the first year after the storm as plant-eating bats declined in number. Similar dramatic changes were documented at Culebrones Cave near Arecibo.

Echolocation calls of the Antillean ghost-faced bat consist of short-duration, frequency-modulated signals that sweep from 63 to 45 kilohertz (Schnitzler et al., 1991). The change in frequency, however, is not consistent; instead, frequency declines slowly at the beginning of the call and more rapidly at the end, in a manner similar to that described for its mainland relative Peter's ghost-faced bat (O'Farrell and Miller, 1999). Each pulse contains several harmonics, and the second of these is the most intense.

Antillean ghost-faced bats typically give birth only once each year to a single offspring (Silva-Taboada, 1979). On Cuba, pregnant females are found from March to June. Ninety percent of Cuban females are pregnant in May, and during this month, adult males and females generally begin roosting in different caves. Births begin in mid-June, and lactating bats are found into September. Neonates are essentially naked, except for a few bristles and vibrissae, and weigh up to 29 percent of the mother's mass. Although a mother may carry her newborn while flying in the roost, especially when disturbed, the young bats are left behind during nightly foraging flights. Ghost-faced bats are in nonbreeding condition from October through at least December, and during this time, males and females typically roost together (Goodwin, 1970; Silva-Taboada, 1979).

Parasites of this species are known best from Cuba (Lancaster and Kalko, 1996; Silva-Taboada, 1979), but ghost-faced bats on Puerto Rico presumably harbor the same or similar organisms. Internal parasites recovered from the ghost-faced bat include flukes, tapeworms, and roundworms. External parasites are a labidocarpid mite, a spinturnicid mite, and two species of argasid tick.

Pteronotus parnellii Gray, 1843
Parnell's Mustached Bat
Murciélago Bigotudo Mayor

Taxonomy: The genera *Mormoops* and *Pteronotus* are the only members of the New World family Mormoopidae. Within *Pteronotus* there are six extant species, and within *Pteronotus parnellii* there are nine living subspecies (Simmons and Conway, 2001). The subspecies found on Puerto Rico is *P. p. portoricensis.* The type locality for *P. parnellii* is an unspecified place on Jamaica, whereas the description of *P. p. portoricensis* is based on a specimen taken from Cueva di Fari, Pueblo Viejo, Guaynabo, Puerto Rico.

Name: The generic name is a combination of the Greek words *pteron,* meaning "wing," and *otus,* which means "pertaining to" (Rodríguez-Durán and Kunz, 1992). Presumably this combination refers to the way in which the wings attach to the body of Davy's naked-backed bat, *Pteronotus davyi,* which was the first member of the genus to be described in the scientific literature. In that naked-backed bat, wing membranes attach to the body at the dorsal midline rather than at the side of the animal; hence the hairless wings completely cover the furred back and give that bat the outward appearance of being naked. The specific epithet *parnellii* refers to Richard Parnell, who obtained the first specimens of *P. parnellii,* and *portoricensis,* of course, refers to the island of Puerto Rico, where the subspecies is found.

Distribution and Status: Parnell's mustached bat is distributed widely in the Neotropics (Map 7). Five of the eight subspecies are found primarily on the mainland, and their collective range extends from Sonora and Tamaulipas in northern Mexico into northern South America east of the Andes and south into the Mato Grosso of Brazil. The remaining subspecies are restricted to the Greater Antilles and associated islands, with *P. p. parnellii* occurring on Cuba and Jamaica, *P. p. pusillus* on Hispaniola, and *P. p. portoricensis* on Puerto Rico (Simmons and Conway, 2001). Biologists recently captured this mormoopid for the first time on St. Vincent in the Lesser Antilles, but identity of the subspecies is not yet known (Vaughan and Hill, 1996).

Parnell's mustached bat is captured readily throughout Puerto Rico,

Fig. 18 Portrait of a Parnell's mustached bat.

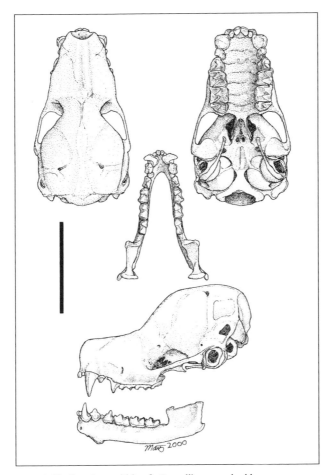

Fig. 19 Skull and mandible of a Parnell's mustached bat
(bar = 10 mm).

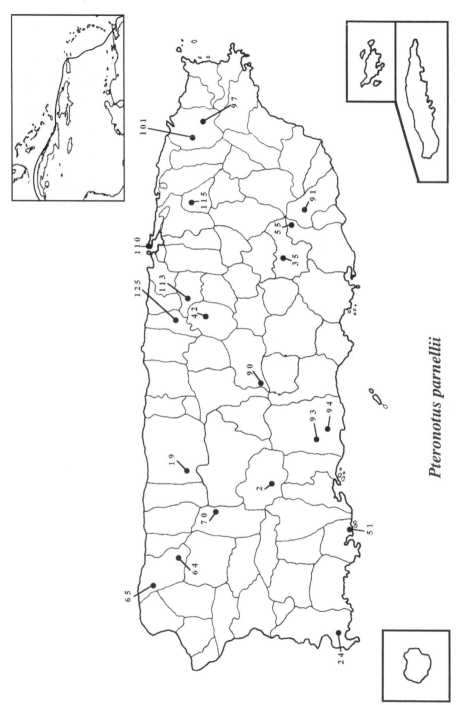

Map 7 Distribution of Parnell's mustached bat on Puerto Rico and in the Caribbean basin. Location of numbered sites is detailed in Appendix 8.

Pteronotus parnellii

although it occurs in low numbers and is the least abundant mormoopid on the island. Its status as an uncommon species on Puerto Rico is not a new phenomenon but was noted by Anthony (1925) more than 80 years ago.

Measurements and Dental Formula: Total length is 79–91 mm; length of tail, 13–22 mm; length of hind foot, 8–13 mm; height of ear, 16–25 mm; and length of forearm, 50–53 mm. Body mass is 10–18 g. Dental formula is incisors 2/2, canines 1/1, premolars 2/3, molars 3/3 = 34.

Description: Parnell's mustached bat is not large, weighing only 10–14 grams, but it is the largest mormoopid on Puerto Rico. The short, broad snout lacks a terminal nose-leaf, although a conspicuous fleshy bump occurs at the base of the muzzle (Fig. 18). The nostrils appear slightly turned up, and tufts of hair protrude from the sides of the snout and give the animal its common name, the mustached bat. The lower lip is slightly warty and has cutaneous flaps that give the impression of a double lip; this double lip is most apparent when a captured animal vocalizes as it is held in the hand. Ears are pointed, narrow, and moderately long, so that when gently laid forward, they extend to the tip of the snout but not much beyond.

The tragus is less ornamented than in other mormoopids, but it still has a small secondary fold on its cranial edge. As in most insectivorous bats, eyes appear small, and the tail membrane is broad. Flight membranes and ears are dark and naked. Although fulvous or orange fur occurs in some parts of the range of this species (Smith, 1972), we have not observed it on Puerto Rico, and it apparently does not occur on Cuba (Silva-Taboada, 1979). Color of fur ranges from dark gray to grayish brown in Puerto Rican specimens.

Cheek teeth of Parnell's mustached bat bear the distinctive W-shaped ridges that are typical of insect-eating species. The skull differs, however, from that of most other insectivorous bats on Puerto Rico in having a palate that has neither incisive foramina nor a gap between the incisors (Fig. 19). These missing features, combined with presence of two upper incisors in each quadrant and a greatest length of about 20–21 millimeters, are useful in distinguishing the skull of Parnell's mustached bat from that of other Puerto Rican species.

Natural History: Parnell's mustached bats roost in humid, warm caves, where these mormoopids often form clusters, sometimes with individuals of other species of bat. In Puerto Rican hot caves, this species is associated most often with brown flower bats and less commonly with long-tongued bats, ghost-faced bats, and sooty mustached bats (Rodríguez-Durán, 1998). Colonies of Parnell's mustached bat typically contain only a few hundred to a few thousand individuals, and such colonies are not as common as those of other species. For example, this mormoopid is found in only 15 percent of the Puerto Rican caves examined by Rodríguez-Durán (1998)—fewer locales than for any other species of bat that is restricted to caves. Its presence, however, is somewhat unpredictable, and one may not catch these bats with certainty at any particular cave.

In Mexico, this mustached bat leaves the cave early each night and appears to follow distinct flyways between the roost and distant foraging areas, with the bats often flying quite low to the ground (Bateman and Vaughan, 1974). On Cuba, the species does not show a well-defined exodus from the dayroost, but it also begins activity early in the evening, from 10 to 35 minutes after sunset (Silva-Taboada, 1979). Our observations in Puerto Rico are similar and suggest that this mormoopid leaves the cave near sunset, typically departing after the sooty mustached bat but before the ghost-faced bat. Large numbers of Parnell's mustached bat do not begin returning to the dayroost until seven hours after the onset of activity, suggesting that the bats spend most of the night outside the cave (Bateman and Vaughan, 1974).

Parnell's mustached bat is a fast-flying species that captures its prey in the air. Although diet of this animal has not been studied on Puerto Rico, feeding habits of some insectivorous species on the island, such as the Brazilian free-tailed bat, appear similar to those of bats on Cuba (Rodríguez-Durán et al., 1993; Silva-Taboada, 1979; Whitaker and Rodríguez-Durán, 1999), and the same may be true of Parnell's mustached bat. On Cuba, this mormoopid feeds mainly on moths, with 93 percent of bats examined including these insects in their diet. Other common types of prey are flies, beetles, ants, orthopterans (grasshoppers and their relatives), and even odonates (dragonflies and damselflies). It may seem contradictory that nocturnal bats consistently are consuming diurnal, visually oriented insects such as odonates, but their presence in the diet probably reflects emergence of the bats from the dayroost at a time when these carnivorous insects are still foraging in the fading light.

These bats typically give birth to a single young, once each year. Timing of reproduction in the Greater Antilles appears variable. Dávalos and Erickson (2003:142) report that 90 percent of females from a cave in Jamaica were "heavily pregnant" in late March. On Cuba, most females are pregnant in June, although at least a few pregnant bats can be found any time from March through July (Silva-Taboada, 1979), and on Puerto Rico, we have captured pregnant females as late as August. Births typically occur in June or July on Cuba, and most mothers are still lactating in September (Silva-Taboada, 1979). Young are born naked and helpless, with their eyes still closed (Herd, 1983). As in other mormoopids, the sexes are segregated in different roosts during the breeding season. For example, at a cave in Mexico during July, 64 females were captured for every male that was handled (Bateman and Vaughan, 1974).

The echolocation system of Parnell's mustached bat is probably the best studied of any mammal (Herd, 1983; Lancaster et al., 1995; Pollak and Henson, 1973; Pollak et al., 1972). Ultrasonic pulses are emitted through the mouth. As the bat flies, emission of sound is synchronized with the wingbeats so that internal pressures unavoidably produced by contracting flight muscles are used to generate the intense sounds. Each pulse consists of two brief, frequency-modulated sweeps of 2–3 milliseconds—one at the beginning and another at the end of the pulse—with a much longer component, 16–28 milliseconds, of constant frequency (about 62 kilohertz) sandwiched between them.

The constant-frequency portion of the call and its resulting echo may help locate and determine the speed of a flying insect relative to the bat by means of a Doppler shift, whereas frequency-modulated segments of the call provide more information about the position and surface details of the prey. Although many bats include short constant-frequency components in their echolocation calls, the long duration of the constant-frequency portion in Parnell's mustached bat calls is unusual among New World species, and the echolocation strategy of this mormoopid is more similar to that of horseshoe bats in the Old World family Rhinolophidae.

The only endoparasites reported are flukes and roundworms from Cuba (Silva-Taboada, 1979). On Cuba, Parnell's mustached bat also harbors streblid batflies and at least three kinds of mites, including one labidocarpid, one myobiid, and one spinturnicid. Another spinturnicid (*Cameronieta thomasi*) is the only parasite of this bat known from Puerto Rico (Gannon and Willig, 1994b). The morphological distinctiveness of parasites such as *C. thomasi* and their restriction to just one or a few mormoopid species supports the placement of mormoopids into a family of their own (Smith, 1972).

Pteronotus quadridens (Gundlach, 1840)
Sooty Mustached Bat
Murciélago Bigotudo Menor

Taxonomy: Bats from 12 different genera currently live on Puerto Rico, but only the mormoopid genus *Pteronotus* is represented by two different species—*P. parnellii* and *P. quadridens.* There are only two subspecies of the sooty mustached bat, *P. q. quadridens* and *P. q. fuliginosus.* The subspecies differ primarily in size, and it is the larger of the two, *P. q. fuliginosus,* that lives on Puerto Rico. *P. quadridens* was described originally from a specimen taken in Canimar, Cuba, whereas the type locality for *P. q. fuliginosus* is Haiti (Rodríguez-Durán and Kunz, 1992).

Name: The generic name *Pteronotus* comes from the Greek words *pteron* and *otus* (Rodríguez-Durán and Kunz, 1992). Together, these words mean "pertaining to the wing," which probably alludes to the unusual wing attachment in Davy's naked-backed bat, *Pteronotus davyi,* the first species placed into the genus. The specific epithet *quadridens* is derived from the Latin *quatri,* meaning "four," and *dens,* meaning "tooth." This combination is a reference to toothlike projections or serrations that occur on the anterior edge of the pinna. The type specimen has four such serrations (Silva-Taboada, 1976), although only two or three occur in many Puerto Rican specimens. The word *fuliginosus* comes from the Latin word for soot and most likely refers to the grayish brown color of many individuals (Jaeger, 1955).

The sooty mustached bat was first described by Gundlach (1840) and named *Lobostoma quadridens,* based on a single specimen captured in Cuba in 1839 and housed at the Museum of Zoology, Humboldt

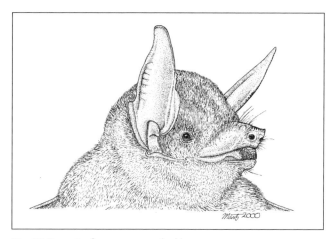

Fig. 20 Portrait of a sooty mustached bat.

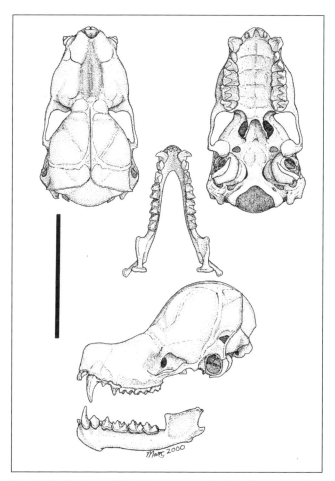

Fig. 21 Skull and mandible of a sooty mustached bat (bar = 10 mm).

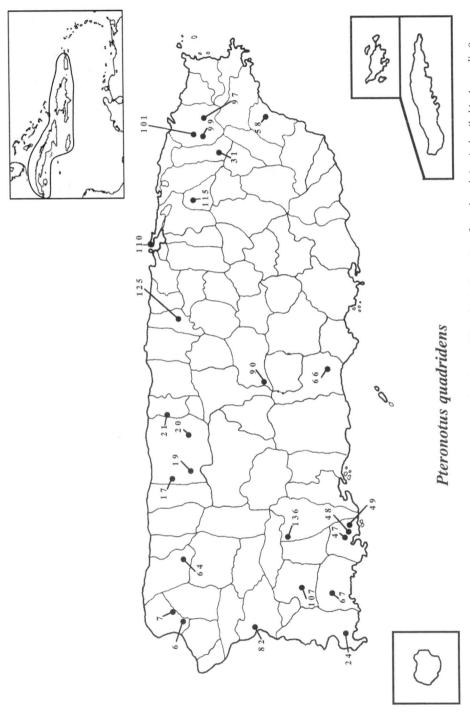

Pteronotus quadridens

Map 8 Distribution of the sooty mustached bat on Puerto Rico and in the Caribbean basin. Location of numbered sites is detailed in Appendix 8.

University, in Berlin. However, Gray (1843), using specimens from Haiti, described an apparently new species that he named *Chilonycteris fuliginosus,* and it is this name that appears in much of the literature pertaining to the sooty mustached bat prior to the 1970s. Smith (1972) ultimately showed that species included in *Chilonycteris* were not sufficiently distinct to be placed in their own genus and that they should be included within the genus *Pteronotus,* a name that pre-dates *Lobostoma* by two years and *Chilonycteris* by one year. Finally, Silva-Taboada (1976) realized that *fuliginosus* and *quadridens* actually refer to the same taxon, and because *quadridens* is older, the correct name for the species is *Pteronotus quadridens.*

Distribution and Status: Unlike Parnell's mustached bat, the sooty mustached bat is found only in the Greater Antilles (Map 8). Although *P. q. quadridens* is restricted to Cuba, *P. q. fuliginosus* is known from Jamaica, Hispaniola, and Puerto Rico. This mammal is abundant on Puerto Rico, and foraging individuals can be captured in most parts of the island.

Like the Antillean ghost-faced bat, the sooty mustached bat is listed as a species of "low risk" by the IUCN (Hutson et al., 2001). Nevertheless, this organization considers both mormoopids as "near threatened," which means that there is a high risk of extinction in the medium-term future and that there is no management program aimed at conservation of either species. This classification reflects the vulnerability of these species as island endemics and their tendency to concentrate in large colonies in few specialized sites, namely hot caves.

Measurements and Dental Formula: Total length is 59–80 mm; length of tail, 13–21 mm; length of hind foot, 6–11 mm; height of ear, 13–19 mm; and length of forearm, 37–40 mm. Body mass is 4–7 g. Dental formula is incisors 2/2, canines 1/1, premolars 2/3, molars 3/3 = 34.

Description: The sooty mustached bat is the smallest bat on Puerto Rico, weighing the equivalent of just two or three pennies. The snout is short and lacks a nose-leaf, but the area around the nostril and lower lip is fairly complex, with cutaneous flaps and wartlike tubercles (Figs. 5, 20). Coarse hairs protruding from the side of the snout form the "mustache," but it is not as evident as in the larger-bodied Parnell's mustached bat. Ears are pointed, narrow, and long so that when laid forward, the ears project obviously beyond the snout. There are two to four tiny toothlike projections along the leading edge of the external ear, and as in most mormoopids, there is a shelflike, secondary fold on the cranial edge of the tragus, although one may need a magnifying glass to see either feature clearly on such a small bat.

Body hairs are distinctly tricolored: dark at the base and tips and a light sandy color in the middle. Overall, the body typically appears grayish brown, but fur of many individuals has a hint of yellow or orange. Ears, wings, and tail membrane, in contrast, are dark and hairless. As in other mormoopids, the terminal 40–50 percent of the tail protrudes from the top of the interfemoral membrane when the animal is at rest, but the entire tail slides into a sheath within the membrane when the legs are spread. The sooty mustached bat is quite similar to Parnell's mustached bat; nevertheless, the sooty mustached bat is distinguished easily by its smaller body size and the presence of serrations on the external ear.

The skull of a sooty mustached bat is most similar to that of Parnell's mustached bat. In both species, the cheek teeth posses W-shaped ridges, the palate lacks incisive foramina, and no gap occurs between the upper incisors (Fig. 21). However, the skull of a sooty mustached bat is considerably smaller, with a greatest length of only 14–15 millimeters.

Natural History: During the day, sooty mustached bats occur exclusively in deep recesses of hot caves, where they form roosting aggregations containing thousands of individuals (Rodríguez-Durán and Lewis, 1987). For example, over 140,000 of these tiny bats roost in Cucaracha Cave south of Aguadilla in the Cordillera Jaicoa. A cave occupied by the sooty mustached bat typically shelters two to five other species, such as the Antillean ghost-faced bat and Greater Antillean long-tongued bat. The different species maintain spatial separation within the cave, but their combined numbers likely contribute to the high temperature and humidity of the caves that they inhabit (Rodríguez-Durán, 1995, 1998; Silva-Taboada, 1979).

Mustached bats roost in sites that allow entrapment of metabolic heat, thus reducing the amount of energy required for thermoregulation and allowing them to have a lowered metabolic rate (Rodríguez-Durán, 1995). For example, basal metabolic rate of the sooty mustached bat is only 55 percent of the value predicted for a terrestrial mammal of comparable size. Other species of bat that use hot caves occasionally roost in cooler caves as clusters or in a lethargic state; sooty mustached bats, however, consistently select high temperatures, both in the field and in the laboratory (up to 39°C; Rodríguez-Durán and Soto-Centeno, 2003).

Temporal separation characterizes the pattern of departure from hot caves that contain multiple species of bat (Rodríguez-Durán and Lewis, 1985, 1987). The sooty mustached bat is the first species to leave the roost each evening, beginning to depart between 11 minutes before and 10 minutes after sunset. These bats exit in a well-formed column that consistently follows the same path, often following a natural corridor across the countryside. They also can fly very high, still maintaining their column, as they move over valleys from one mountainous area to another; for example, columns of these bats leave Cucaracha Cave and fly high across the valley of the Río Culebrinas to the hills of La Cadena, 9 kilometers away. Although some activity occurs at the cave throughout the night, most individuals do not return until dawn, with the majority arriving between 17 minutes before and 10 minutes after sunrise.

Hot caves are not common in the Puerto Rican landscape (Rodríguez-Durán, 1998), and capture of sooty mustached bats throughout the island indicates an ability to travel long distances. As already noted, bats from Cucaracha Cave fly in columns for over 9 kilometers before dispersing to their foraging areas, but how far individuals travel after the groups break up is unknown (Rodríguez-Durán, 1984). In Cuba, sooty mustached bats are capable of returning to their home roost by dawn after being released 30 kilometers away earlier in the night, indicating a good homing sense (Silva-Taboada, 1979). Flight speed while commuting is about 27 kilometers per hour (Rodríguez-Durán and Lewis, 1985).

The early evening departure of the sooty mustached bat exposes this mammal to predation by diurnal predators, especially raptors such as the red-tailed hawk, American kestrel, and merlin (Rodríguez-Durán and Kunz, 1992; Rodríguez-Durán and Lewis, 1985). The birds typically capture these bats in midair, kill them in flight, and then carry the body to a perch, where the predator removes the wings and eats the remainder. In one study at Cucaracha Cave, up to nine merlins hunted these bats on any given day, with each bird capturing an average of three bats per night; an estimated 3,100–4,700 bats could be consumed in this manner during a year (Rodríguez-Durán and Lewis, 1985). Feral house cats and snakes also capture exiting bats, including some sooty mustached bats.

Like many insectivorous species of bat, the sooty mustached bat is probably an opportunistic forager to some degree. Insects from one to seven different orders have been found in the stomach or fecal sample of a single bat. Although sooty mustached bats do not have a particularly robust skull, they rely heavily on hard-bodied prey, particularly beetles (Rodríguez-Durán et al., 1993). Moths, flies, and true bugs are taken consistently as well, and wasps and flying ants, which often occur in large but unpredictable swarms, are eaten when available. Flying insects are detected using echolocation calls that consist of a very short pulse of constant frequency that turns into a longer, frequency-modulated sweep from about 84 to 68 kilohertz (Macías and Mora, 2003; Schnitzler et al., 1991).

Females come into heat just once each year and generally give birth to a single, naked youngster; twinning is extremely rare (Silva-Taboada, 1979). Pregnant females are found from February through June, with the largest percentage occurring in May, when births begin. Although most females are lactating in July, some young still may be suckling into September (Rodríguez-Durán, 1984; Silva-Taboada, 1979). A neonate is quite large relative to the mother, and its weight may exceed 30 percent of the mother's body mass. Not surprisingly, a mother does not carry her newborn on foraging bouts, but leaves it behind in a dense cluster of 50–200 other young bats until she returns. At many but not all caves, sexual segregation occurs, with either adult males or females disappearing from established roosts early in the breeding season and returning later in the year.

As one might expect of a species living in such large assemblages, the sooty mustached bat hosts a variety of ectoparasites (Rodríguez-Durán and Kunz, 1992). These include at least six species of mite, two ticks, and two streblid batflies. This bat also is infested with a variety of flukes, tapeworms, roundworms, and spiny-headed worms. Despite the abundance of this bat on Puerto Rico, all documented parasites are from Cuba.

Family Phyllostomidae
American Leaf-nosed Bats

The Phyllostomidae is one of six bat families that occur only in the Western Hemisphere. Although phyllostomids primarily live in tropical and sub-tropical regions, the family is widespread. These bats range from the south-western United States to northern Argentina and central Chile, and they are found throughout the West Indies. The family is large, containing 7 subfamilies, 54 genera, and 162 species, or about 15 percent of all species of bat worldwide (Simmons, 2005; Wetterer et al., 2000). Five species of phyllostomids, representing three subfamilies, live on Puerto Rico.

The family is named for a leaflike appendage called a nose-leaf that projects above and often to the sides of the paired nostrils. Although some bats, such as the mormoopids and vespertilionids, emit echolocation sounds through an opened mouth, other bats, including the phyllostomids, emit orientation sounds through the nose. The function of the nose-leaf is not understood totally, but it seems to help focus sound waves into a nar-row beam as they leave the adjacent nostrils (Hartley and Suthers, 1987). Echolocation sounds produced by leaf-nosed bats are similar to those used by most other bats, except that those produced by phyllostomids are quite low in intensity; that is, the sounds are not very loud. Consequently, phyl-lostomids are somewhat difficult to detect with ultrasonic bat detectors, and these animals often are referred to as "whispering bats."

The family Phyllostomidae includes many small, nectar-feeding species that weigh only 5–10 grams, but it also contains the largest bat in the New World, the false vampire bat, with a body mass up to 225 grams and a wingspan approaching 1 meter. Color of fur ranges from white in the ghost bats of Central America to various shades of red, gray, brown, or black in other phyllostomids. Although most species are evenly colored, facial or body stripes occur in some (e.g., white-lined bats).

Members of the Phyllostomidae vary greatly in behavior and ecology (Fenton and Kunz, 1977). For example, some phyllostomids gather in colonies containing hundreds or thousands of animals, whereas others are solitary. Mating systems range from monogamy in the false vampire to polygyny ("harems") in the greater spear-nosed bat and many others. Some leaf-nosed bats find shelter in caves, buildings, or hollow trees, whereas others simply roost among the leaves. The tent-making bat is one of the few bats in the New World that creates its own individualized shelter. These bats manufacture "tents" by using their teeth to cut veins and/or midribs of a leaf so that portions of the leaf collapse around the bat, thus protecting the animal from sun, rain, and the eyes of hungry predators (Kunz et al., 1994). Diet, like shelter, also is highly variable, and different species of phyllostomid consume insects, leaves, fruit, pollen, nectar, blood, or small vertebrates, including birds, rodents, frogs, lizards, and even other bats (Gardner, 1977).

Artibeus jamaicensis Leach, 1821
Jamaican Fruit Bat
Murciélago Frutero Común

Taxonomy: The genus *Artibeus* belongs to the fruit-eating subfamily Stenodermatinae—the largest phyllostomid subfamily, with up to 17 genera and 69 species (Simmons, 2005). Depending on the authority followed, number of species within the genus ranges from about 10 to perhaps more than 20. Much of the controversy centers on whether smaller species that traditionally were included in *Artibeus* should be placed in separate genera (Nowak, 1999; Owen, 1988; Van Den Bussche et al., 1993). In any event, *Artibeus jamaicensis* is the only member of the genus found on Puerto Rico and in the rest of the Greater Antilles. There may be up to 12 subspecies of *A. jamaicensis* (Hall, 1981; Silva-Taboada, 1979; Simmons, 2005), but all individuals from the Antilles, except those in Cuba and some small islands north of Hispaniola, belong to the subspecies *Artibeus jamaicensis jamaicensis.*

Name: The generic name *Artibeus* may be derived from two Greek words referring to facial stripes that occasionally are present (Ortega and Castro-Arellano, 2001). The epithet *jamaicensis* refers to the island of Jamaica, which is the type locality of the species and subspecies. This bat is abundant throughout the Neotropics, and consequently it is sometimes called the common fruit bat. Wilson and Cole (2000) prefer the name Jamaican fruit-eating bat.

Distribution and Status: The Jamaican fruit bat has a broad distribution throughout Central and most of South America, as well as on almost all Caribbean islands and even the Florida Keys (Ortega and Castro-

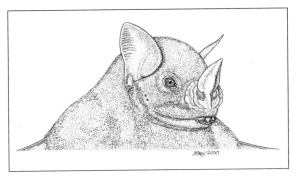

Fig. 22 Portrait of a Jamaican fruit bat.

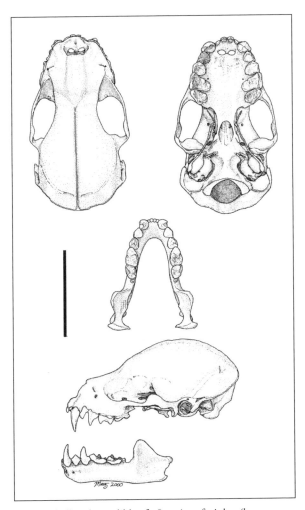

Fig. 23 Skull and mandible of a Jamaican fruit bat (bar = 10 mm).

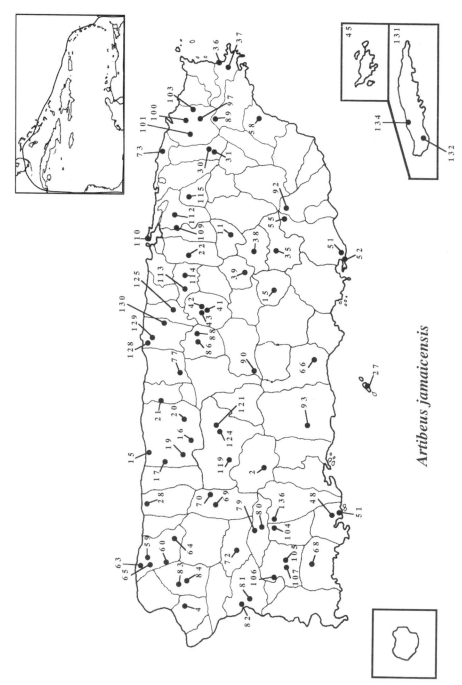

Artibeus jamaicensis

Map 9 Distribution of the Jamaican fruit bat on Puerto Rico and in the Caribbean basin. Location of numbered sites is detailed in Appendix 8.

Arellano, 2001; Map 9). On Puerto Rico this species is moderately abundant and commonly encountered, being found in virtually all habitats. The Jamaican fruit bat is also known from the nearby islands of Culebra, Vieques, Caja de Muertos, and Mona.

Measurements and Dental Formula: Total length is 81–89 mm; length of hind foot, 15–18 mm; height of ear, 20–24 mm; and length of forearm, 58–64 mm. Body mass is 40–53 g. Dental formula is incisors 2/2, canines 1/1, premolars 2/2, molars 2/2 = 30.

Description: The Jamaican fruit bat is a large bat with short, velvety fur. Overall, body color ranges from dull brown to gray or black with a silvery tinge; underparts are usually paler. As in all stenodermatines, there is a well-developed nose-leaf (Fig. 22) but no external tail. Wings are short and broad (i.e., have a low aspect ratio), and the wing and tail membranes are black, somewhat thick, and tough. The tail membrane is narrow, appearing U-shaped when the legs are spread, and there is a short but distinct calcar (Fig. 8). The low aspect ratio and small tail membrane presumably facilitate carrying food and moving through tangled vegetation, respectively. White facial stripes are present above and below the eye, although striping varies from distinct to faint and may be absent.

Large size (greatest length over 27 millimeters) and presence of incisive foramina separate the skull of the Jamaican fruit bat (Fig. 23) from that of all other species, except the Antillean fruit bat. The Jamaican fruit bat, however, has only four upper cheek teeth on each side, compared with five for the Antillean fruit bat.

Natural History: Throughout its range, the Jamaican fruit bat consumes a variety of plant materials—fruits, flowers, leaves, and nectar—and occasionally insects (Gardner, 1977). Although figs are the most common food in many parts of the New World, Willig and Gannon (1996) report that 65 percent of 40 bats from the Luquillo Mountains for which stomach contents were examined had fed on fruit from the trumpet tree, with less than 5 percent of the bats consuming fruit of bulletwood, elder, or sierra palm. In contrast, Kunz and Diaz (1995) found seeds of turkey berry in 56 percent of fecal pellets gathered in Aguas Buenas Cave, along with seeds of jag (51 percent) and trumpet tree (13 percent). Almond and elder were the two most common fruits detected in feces of Jamaican fruit bats at Convento Cave, in Corozal, although almond almost disappeared from the diet in the year following Hurricane Georges (Rodríguez-Durán and Vázquez, 2001).

We have seen these bats eating mango, and others have reported angelin, golden apple, golden-fruited palm, hogplum, maría, papaya, and pitch apple as foods on Puerto Rico (Díaz-Díaz, 1983; Kunz and Díaz, 1995; Rodríguez-Durán and Vázquez, 2001). Included in the diet are leaves of figs, maría, and mountain immortelle; leaves apparently provide a more concentrated source of protein than do fruits (Kunz and Diaz, 1995). These bats also occasionally lap drops of nectar that hang from flowers of banana plants (Rodríguez-Durán and Kunz, 2001).

On Puerto Rico, the Jamaican fruit bat often spends the day in caves (Fig. 11), ranging from some that are shallow and well lit to those that are deep and totally dark, with little preference for one type over another (Anthony, 1918, 1925). Consequently, this abundant bat is found in a greater proportion of caves (66 percent) on Puerto Rico than any other species (Rodríguez-Durán, 1998). Jamaican fruit bats on Vieques, where caves are scarce, have adopted abandoned military bunkers made of concrete as roosting sites, with occupied bunkers apparently more isolated (protected from disturbance) than those that are unoccupied (Rodríguez-Durán, 2002). In addition to cavelike structures, the Jamaican fruit bat occasionally roosts in buildings and vegetation (Díaz-Díaz, 1983). Hollow trees and leaf tents also are important roosts in other parts of this bat's range (Handley et al., 1991; Kunz and Lumsden, 2003), but their use has not been documented on Puerto Rico.

In Panama, the Jamaican fruit bat usually begins foraging about 30 minutes after sunset, and on dark nights it may continue feeding until 30 minutes before sunrise. However, on moonlit nights, it returns to the dayroost for several hours during the brightest part of the night. Morrison (1978c) speculates that such "lunar phobia" is a means of avoiding predation by nocturnal hunters, such as owls. Avian predators capable of feeding on bats are absent from Puerto Rico and, as one might predict, Jamaican fruit bats do not display lunar phobia on the island (Rodríguez-Durán and Vázquez, 2001).

Ripe fruit is detected from a distance by its odor, and the bat uses echolocation only to avoid obstacles and perhaps to aid in the final "capture" (Kalko et al., 1996; Thies et al., 1998). This mammal typically does not eat in the tree in which the fruit was growing. Instead, the animal flies 25–200 meters from the fruiting tree to a feeding roost located in a different tree or perhaps in the entrance to a cave. Here the bat chews pieces of fruit or leaves, swallows the juice, and spits out any remaining fibrous material in the form of a chewed pellet (Handley et al., 1991; Kunz and Diaz, 1995). Repeated use of the same feeding roost leads to an accumulation of pellets beneath the roost (Fig. 12).

In Panama, the Jamaican fruit bat returned to the same fruiting tree for up to eight consecutive nights (Handley et al., 1991; Morrison, 1978a), but on Puerto Rico this species did not exhibit such short-term fidelity. Over a six-week period we radiotracked more than 20 individuals and found that these bats did not return to, or even near, foraging areas where they were first captured in the Luquillo Mountains. Bats were extremely difficult to locate after initial release, and in fact, day-roosts never were located, indicating that they were a long distance from the foraging area—at least 10 kilometers. Such large commuting distances suggested that density of fruiting trees was rather low (Morrison, 1978b). Although radiotagged individuals did not exhibit short-term fidelity to foraging sites, we occasionally found banded individuals in the same area of the Luquillo Experimental Forest after one year or more.

In September 1989, the eye of Hurricane Hugo passed within 10 kilometers of the Luquillo Experimental Forest, and afterward, the number of Jamaican fruit bats captured in the forest greatly decreased (Gannon and Willig, 1994a). Because this mammal is a strong flier and capable of commuting long distances, the reduced number of individuals after the hurricane may reflect movement of bats to less severely affected areas of the island. This species was virtually absent from the forest following the hurricane, but after two years, the population returned to prehurricane levels, presumably in response to recovering vegetation and increased availability of fruit. Several individuals captured in the Luquillo Experimental Forest at this time had been banded there prior to the hurricane, which supports the idea that the bats focused their activity elsewhere on the island immediately after the storm. The absence of Jamaican fruit bats following Hurricane Hugo indicates that this species plays a minor role in seed dispersal of early successional plants after a major disturbance, such as a hurricane.

Early reports suggested that the Jamaican fruit bat was not highly social (Anthony, 1918, 1925). However, more recent observations at Aguas Buenas Cave show that bats of this species form harems during the reproductive period (Kunz et al., 1983). A single adult male defends a group of 2–14 reproductive females that cluster together inside a solution cavity in the ceiling of a cave. Larger, older males are better able to defend groups of females and hence are more likely to be "harem masters" than are smaller, younger males. Young males, not yet in breeding condition, occasionally cluster with reproductive females or roost by themselves in another area of the cave.

Reproductive patterns in this mammal apparently vary throughout its range, but within the Caribbean basin, most adult females give birth twice per year (polyestry), with parturition occurring during two specific times of the year (bimodal polyestry), coinciding with the rainy season when fruits are abundant. For example, most young are born in March–April and July–August in Panama and Cuba (Handley et al., 1991; Silva-Taboada, 1979). Our unpublished records and those from the literature (Anthony, 1925; Gannon and Willig, 1992; Kunz et al., 1983) indicate that pregnant females are found on Puerto Rico primarily during January–March and May–July, with lactating females being reported from March to July. Data from the rest of the year are sparse, and frequency of breeding late in the dry season still needs to be determined for this species on Puerto Rico.

Jamaican fruit bats typically mate shortly after giving birth (postpartum estrus); hence it is common for females to be both pregnant and lactating at the same time. Litter size is only one. Neonates are fully furred, have open eyes, and weigh more than 25 percent of the mother's nonpregnant mass (Handley et al., 1991). Gestation lasts 3.5–4 months for females mated early in the year, but in some parts of the bat's range, the second pregnancy involves "delayed development." After implanting in the wall of the uterus, the embryo becomes quiescent, delaying resumption of cell division for up to two months. Such a delay presumably allows young to be born during a time of food abundance rather than during the dry season when fruits are less available (Fleming, 1971). Whether such a delay occurs in some Jamaican fruit bats on Puerto Rico is not known. Life span in the wild is up to nine years (Handley et al., 1991).

Like most mammals, this species harbors a variety of ectoparasites (Webb and Loomis, 1977). In the Luquillo Experimental Forest, 87 percent of Jamaican fruit bats are infested with an average of six parasites per host (Gannon and Willig, 1994b). Although level of infestation does not vary throughout the year, juvenile bats consistently harbor a greater number of ectoparasites than do adults. Parasites include four species of streblid batfly (*Aspidoptera phyllostomatus, Megistopoda aranea, Trichobius intermedius,* and *T. robynae*); two spinturnicid wing mites (*Periglischrus iheringi* and *P. vargasi*); two labidocarpid mites (*Paralabidocarpus artibei* and *P. foxi*); and one spelaeorhynchid ear mite (*Spelaeorhynchus praecursor*). The wing mite, *P. iheringi,* dominates, occurring on 69 percent of the bats. On Cuba this bat is infested with spinturnicid, labidocarpid, and myobiid mites as well as argasid ticks; internal parasites include flukes, tapeworms, and roundworms (Silva-Taboada, 1979).

Brachyphylla cavernarum Gray, 1834
Antillean Fruit Bat
Murciélago Hocíco de Cerdo

Taxonomy: The genus *Brachyphylla* is the sole member of the phyllostomid subfamily Brachyphyllinae (Simmons, 2005). *Brachyphylla* contains only two species—*B. cavernarum* and *B. nana*—both of which are restricted to the Antilles. There are three subspecies within *B. cavernarum,* and the one living on Puerto Rico is *Brachyphylla cavernarum intermedia*. The species *B. cavernarum* was described originally from a specimen taken on St. Vincent in the Lesser Antilles, and the type specimen for the subspecies *B. c. intermedia* was captured near Corozal, Puerto Rico (Swanepoel and Genoways, 1978).

Name: The generic name *Brachyphylla* is formed from two Greek words, *brachys* and *phyllon,* meaning "short leaf," a reference to the bat's greatly reduced nose-leaf. The specific epithet *cavernarum* is Latin and refers to caves where this species often roosts. The subspecific designation indicates that *B. c. intermedia* generally is intermediate in size between the other two subspecies, *B. c. cavernarum* and *B. c. minor*. Some authors refer to this species as the Lesser Antillean fruit bat, although Wilson and Cole (2000) encourage use of the name Antillean fruit-eating bat.

Fig. 24 Portrait of an Antillean fruit bat.

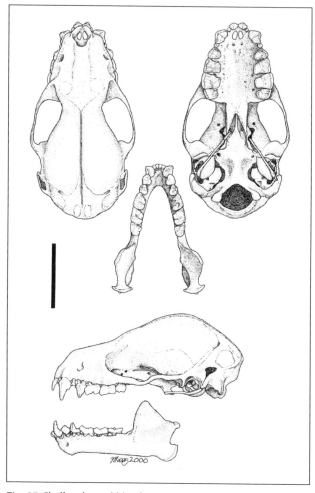

Fig. 25 Skull and mandible of an Antillean fruit bat (bar = 10 mm).

Distribution and Status: *Brachyphylla cavernarum* likely originated in the Greater Antilles and subsequently invaded islands throughout the Lesser Antilles (Baker and Genoways, 1978; Koopman, 1975). Today this fruit bat occurs east of the Mona Passage on Puerto Rico and the Virgin Islands, and on the Lesser Antilles south to St. Vincent and Barbados (Map 10). This phyllostomid occurs on at least 13 of the 19 major islands throughout its range, and it is one of the most common species of fruit-eating bat on Puerto Rico. In addition to Puerto Rico, the subspecies *B. c. intermedia* occurs on the Virgin Islands, with the exception of St. Croix (Swanepoel and Genoways, 1978).

Measurements and Dental Formula: Total length is 85–94 mm; length of hind foot, 17–21 mm; height of ear, 20–24 mm; and length of forearm, 61–68 mm. Body mass is 36–53 g. Dental formula is incisors 2/2, canines 1/1, premolars 2/2, molars 3/3 = 32.

Description: The Antillean fruit bat and Jamaican fruit bat are similar in size and are two of the largest bats on Puerto Rico. These species, however, readily are distinguished from each other. Although the Jamaican fruit bat has a well-formed nose-leaf, that of the Antillean fruit bat is reduced greatly, so much so that a novice might not even consider one to be present (Figs. 5, 24). The nostrils and rudimentary nose-leaf appear at the flattened end of an otherwise conical snout and give the Antillean fruit bat a distinctive piglike appearance. Hairs are usually yellowish white at the base, with tips that are somewhat more brownish; darker patches of fur may occur on the head, neck, or back. Ears are naked and dark brown. The tail is miniscule, only a few millimeters long, and totally enclosed in the base of the tail membrane, making it difficult to detect (Fig. 9). The tail membrane is much smaller than that of some bats, such as vespertilionids or molossids, but similar in development to that of Jamaican fruit bats.

Like those of other phyllostomids, the skull of an Antillean fruit bat possesses well-formed incisive foramina (Fig. 25), but it is quite large, with a greatest length of 30–32 millimeters. The skull is somewhat similar in size and shape to that of the Jamaican fruit bat, but the latter species has only four cheek teeth in each side of the upper jaw, compared with five in the Antillean fruit bat.

Natural History: Despite its abundance, the biology of this bat on Puerto Rico remains poorly known. Individuals have been observed in a vari-

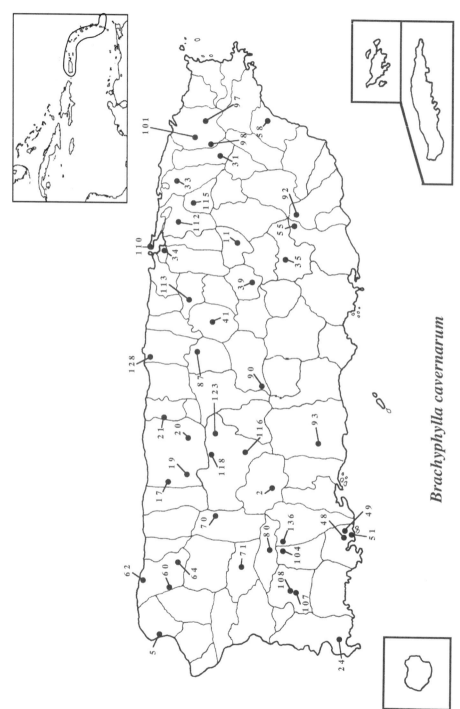

Map 10 Distribution of the Antillean fruit bat on Puerto Rico and in the Caribbean basin. Location of numbered sites is detailed in Appendix 8.

Brachyphylla cavernarum

ety of habitats ranging from the xeric Guánica State Forest to the mesic Luquillo Experimental Forest. Although we often have captured this fruit bat in nets set at ground level, others have netted this species above the forest canopy (Jones et al., 1971).

As its scientific name implies, the Antillean fruit bat usually roosts in caves on Puerto Rico. Its presence, however, in some habitats where caves are absent (e.g., Luquillo Experimental Forest) indicates that it either roosts in other situations or is capable of commuting long distances. On St. Croix, it occasionally occupies buildings or roosts in dense tree canopy (Beatty, 1944; Bond and Seaman, 1958). Nellis and Ehle (1977) describe a colony on St. Croix using a large well (4 m wide and 50 m deep) that was exposed to direct sunlight, but such tolerance of light is not characteristic of the species. Instead, Antillean fruit bats typically choose an area that is dimly lit or totally dark for roosting. Antillean fruit bats in the Guánica State Forest, for example, are always found in the darkest recesses of caves. This species is similar to the Jamaican fruit bat in that both prefer roosting in open, well-ventilated places with cooler temperatures than those in roosting sites typically sought by the mormoopids or Antillean long-tongued bats. Nevertheless, on occasion, Antillean fruit bats roost at temperatures as high as 26°C.

Little is known about the social organization of Antillean fruit bats, other than that they typically form colonies in caves. These bats aggregate in colonies containing up to 2,000 individuals on St. Croix; 6,000 on Nevis; and 10,000 on Cuba (Bond and Seaman, 1958; Pedersen et al., 2003; Silva-Taboada and Pine, 1969). The largest colonies on Puerto Rico appear similar in size. The Antillean fruit bat is seldom the only species roosting at a site, and one may find it occupying the same cave as long-tongued bats, brown flower bats, or other species, although the various species typically do not interact or associate with each other while in the caves (Rodríguez-Durán, 1998).

Conversely, Jamaican and Antillean fruit bats apparently avoid roosting near each other, because the two species generally do not share the same cave despite their similarity in roosting requirements. In any event, an Antillean fruit bat is quite active during the day, continually grooming, quarreling, and moving about but rarely sleeping. Other observers describe this phyllostomid as belligerent and pugnacious

(Anthony, 1918; Swanepoel and Genoways, 1983), and we totally agree.

Diet consists of insects, pollen, and especially fruit (Nellis, 1971; Nellis and Ehle, 1977; Pedersen et al., 2003). Antillean fruit bats forage in the canopy or take fruit that has fallen to the ground. On St. Croix, this phyllostomid consumes mango, papaya, and sapodilla. Captive bats accept bananas, apples, grapes, melons, peaches, and pears but not citrus fruits, and they readily eat flowers from a number of trees, including portia and silk cotton. Other fruits eaten on Puerto Rico are angelin and maría. We have maintained captive individuals from the Luquillo Experimental Forest for about three weeks on a diet of banana, guava, and mango and found that these bats exhibit a strong preference for mango. As in other fruit-eating species, the Antillean fruit bat extracts the juice from pulpy fruits and ejects most fibrous material in the form of a pellet; food that is swallowed may pass through the entire digestive tract in only 15 minutes (Nellis and Ehle, 1977). Although fruit dominates the diet, over 40 percent of the individuals sampled from Puerto Rico had consumed at least some insects, particularly beetles (Soto-Centeno et al., 2001).

On St. Croix, the Antillean fruit bat is aggressive toward Jamaican fruit bats that forage in the same area (Nellis and Ehle, 1977). When Antillean fruit bats begin feeding in a tree, Jamaican fruit bats often move to different parts of the tree or vacate the area. Antillean fruit bats sometimes land near a feeding Jamaican fruit bat and actually chase it through the branches. Although Nellis and Ehle (1977) indicate that the Antillean fruit bat is usually successful at displacing the other phyllostomid from feeding areas where they co-occur, no evidence exists for this on Puerto Rico. Mistnetting in three habitats within the Luquillo Experimental Forest indicates that the Jamaican fruit bat is always present in higher numbers than is the Antillean fruit bat (Gannon and Willig, 1994a), although low numbers of Antillean fruit bats may be related to the lack of caves on this part of the island.

Nothing is known about the mating system of this island endemic, and knowledge of its reproductive patterns is mostly fragmentary. Twelve females from the island of Caicos carried similar-sized fetuses in March (Buden, 1977), and on St. Croix females from one colony gave birth during a three-week period in late May and early June (Bond and

Seaman, 1958; Nellis and Ehle, 1977). A pregnant female was captured near Utuado on 15 March (Timm and Genoways, 2003), and lactating females from Puerto Rico have been observed on 5 July (Anthony, 1918). We frequently have netted females that were simultaneously pregnant and lactating, indicating that the Antillean fruit bat on Puerto Rico is polyestrous and undergoes a postpartum estrus. We have captured such females in January, March, April, June, July, and August. Hence these bats may be capable of breeding throughout the year, although additional netting during autumn is needed to verify this. In any event, youngsters are hairless, weigh about 10–15 grams at birth, and make their first flight at about two months of age (Nowak, 1999).

Only a few ectoparasites are known for this bat (Swanepoel and Genoways, 1983), and no quantitative studies exist on rates of infestation. Documented parasites from Puerto Rico include a macronyssid and spinturnicid mite (*Radfordiella oudemansi* and *Periglischrus cubanus,* respectively); two species of streblid batfly (*Megistopoda aranea* and *Trichobius truncatus*); and two species of labidocarpid mite (*Lawrence-ocarpus micropilus* and *L. puertoricensis;* Anthony, 1918; Gannon and Willig, 1994b).

Erophylla sezekorni Gundlach, 1861
Brown Flower Bat
Murciélago de las Flores

Taxonomy: *Erophylla* and *Phyllonycteris* are the only genera within the phyllostomid subfamily Phyllonycterinae, a small group of bats endemic to the West Indies (Simmons, 2005). Although fossil *Phyllonycteris* are known from Puerto Rico (Anthony, 1925; see Table 5), only *Erophylla* lives on the island today. All West Indian populations belong to a single species, *E. sezekorni* (Buden, 1976). There are two subspecies, *E. s. sezekorni* and *E. s. bombifrons,* and it is the latter taxon that is found on Puerto Rico. The type specimen for *E. sezekorni* is from Cuba, whereas the description of *E. s. bombifrons* is based on a bat that was collected near Bayamón, Puerto Rico. Some taxonomists believe that the two subspecies are distinct species (e.g., Simmons, 2005), so the name *Erophylla bombifrons* occasionally appears in literature on this Puerto Rican bat.

Fig. 26 Portrait of a brown flower bat.

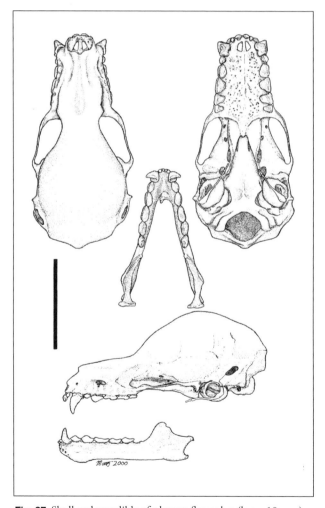

Fig. 27 Skull and mandible of a brown flower bat (bar = 10 mm).

Name: The generic name *Erophylla* comes from the name of the Greek god of love, Eros, and the word for leaf, *phyllon*; the combination refers to the vague similarity of the nose-leaf to the love dart of Eros. The specific epithet honors Geheimerath Sezekorn, who provided Juan Cristobal Gundlach with the holotype. The subspecific epithet *bombifrons* comes from Latin and describes the somewhat inflated braincase of this bat (Baker et al., 1978).

Distribution and Status: This species is endemic to the West Indies and occurs on the Bahama Islands, Caicos Islands, Puerto Rico, Hispaniola, Jamaica, Cayman Islands, and Cuba (Map 11). The subspecies *E. s. bombifrons* is known only from Hispaniola and Puerto Rico. The brown flower bat is common and is found throughout Puerto Rico.

Measurements and Dental Formula: Total length is 80–88 mm; length of tail, 13–17 mm; length of hind foot, 13–15 mm; height of ear, 17–19 mm; and length of forearm, 47–51 mm. Body mass is 16–21 g. Dental formula is incisors 2/2, canines 1/1, premolars 2/2, molars 3/3 = 32.

Description: The brown flower bat is a medium-sized bat with an elongate muzzle (Fig. 26). Its long, pointed tongue is protrusible and coated with many bristlelike papillae that may aid in lapping nectar. The nose-leaf has small extensions lateral to the nostrils but is relatively short compared with that of most phyllostomids. The well-formed tail is shorter than the femur, but it extends past the posterior edge of the narrow, U-shaped uropatagium. There is a small vestigial calcar. Fur is short and silky, and the body appears chestnut brown above and paler below. Body hairs are distinctly bicolored, snow white at the base but brown at the tip. The head and face, in contrast, appear lighter in color, with shorter, unicolored hairs. Flight membranes and ears are light brown and naked.

On Puerto Rico this species is somewhat similar in appearance to the Greater Antillean long-tongued bat, but the brown flower bat has a larger body, thicker muzzle, lighter brown color, and proportionately smaller nose-leaf. In addition, the brown flower bat is often confused with the Antillean fruit bat because of similar coloration, belligerent behavior, and a small nose-leaf, but the Antillean fruit bat is about the twice the size of the brown flower bat.

The skull of a brown flower bat differs from that of most other species on Puerto Rico in having paired incisive foramina and an elon-

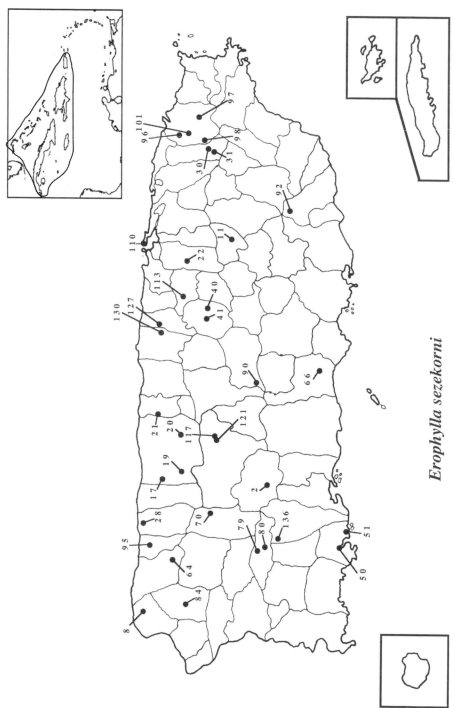

Map 11 Distribution of the brown flower bat on Puerto Rico and in the Caribbean basin. Location of numbered sites is detailed in Appendix 8.

Erophylla sezekorni

gate rostrum (Fig. 27). Its skull is most similar to that of the Greater Antillean long-tongued bat, but that species possesses a pinhole-sized foramen in the midline of the palate, anterior to the incisive foramina, and this is lacking in the brown flower bat. In addition, greatest length of the skull for a brown flower bat (24–25 millimeters) is slightly greater than that of a long-tongued bat (20–21 millimeters).

Natural History: During the day, the brown flower bat typically joins multispecies assemblages in caves but roosts separately from the other species. It normally roosts in large colonies (thousands or hundred thousands of bats) in cooler portions (26–28°C) of "hot caves," and it selects these lower temperatures in laboratory experiments as well (Rodríguez-Durán and Soto-Centeno, 2003). Occasionally groups of a few hundred flower bats occupy cooler caves, where they roost within small cavities in the ceiling or walls. Presumably these cavities facilitate entrapment of metabolic heat, thus reducing the amount of energy required for thermoregulation. Individuals obtained from hot caves have basal metabolic rates that are lower than those of similar-sized fruit- or nectar-feeding bats that use other types of roosts (Rodríguez-Durán, 1995).

Brown flower bats begin foraging later than most other Puerto Rican species of bat. At Culebrones Cave near Arecibo, brown flower bats gather with sooty mustached bats, long-tongued bats, and ghost-faced bats to form a colony of about 300,000 individuals. The tiny mustached bats are the first to leave every evening, departing immediately before or immediately after sunset, but peak exit of brown flower bats does not occur until 40–60 minutes after sunset (Rodríguez-Durán, 1996).

Although this late pattern of activity minimizes contact with diurnal predators, such as hawks, brown flower bats are an important prey of an endemic species of snake, the Puerto Rican boa (Rodriguez and Reagan, 1984; Rodríguez-Durán, 1996). Up to 21 boas gather inside and at the entrance of Culebrones Cave in a single evening. Here, the snakes hang from rocks, logs, or exposed tree roots and dangle their bodies in the space through which the bats are flying as they exit the cave (Fig. 28, Color Plates). When a bat accidentally bumps into the head of a snake, the reptile quickly captures the mammal in its mouth, throws two coils around its prey, and eventually eats the hapless victim; some bats are grabbed after merely caressing the snake with a moving wing or even while hovering in front of the motionless reptile. This type of bat hunting does not have a high rate of success, and individual snakes may go for days without a capture. Although the number of bats eaten by snakes is low relative to the total population of bats, this resource may be critical for the endangered boas.

Prior to Hurricane Georges in 1998, hunting activity of snakes generally coincided with the exit of brown flower bats (Rodríguez-Durán, 1996). The other bat species, which leave earlier, are much smaller than the brown flower bat, and presumably the energetic reward in capturing the smaller animals was not great enough to compensate the boa for its efforts. However, the population of brown flower bats greatly decreased following Hurricane Georges (Jones et al., 2001), and snakes now appear at the entrance earlier each night. Apparently the boas are responding to the decreased abundance of large prey by increasing the number of a smaller-bodied, mustached or ghost-faced bats that they consume.

Diet of the brown flower bat typically contains some combination of insects, nectar, and fruit. In a very detailed study, Soto-Centeno (2004) examined fecal pellets and pollen samples from 109 brown flower bats living in Culebrones Cave and found that 75 percent of the individuals had consumed at least some insect remains, 76 percent had fed on nectar, and 85 percent had eaten fruit. About half the bats had consumed all three dietary items during their most recent foraging period, and 40 percent of the animals used at least two of these foods. Beetles are the type of insect that is most common in the diet, but flies, moths, and flying ants also are eaten. Seeds found in fecal pellets indicate that brown flower bats often feed on the fruit of Panama berry, elder, and turkey berry, whereas pollen samples obtained from the fur suggest that this bat also visits flowers of banana, guava, and wild tamarind for nectar.

The brown flower bat has a litter size of only one and probably gives birth just once each year in spring. On Puerto Rico pregnant females can be obtained from February through June and lactating females from May through September. A newborn weighs about 4.5 grams or 25 percent of the mother's postpartum mass (Soto-Centeno and Kurta, 2003).

At birth, wing membranes lack pigment and are translucent. The body is naked and appears somewhat pinkish, although youngsters quickly develop body hairs that are lighter and more uniformly colored than those of the adults. Eyes open before birth, but ears do not become fully erect until after the young are born. Neonates typically cling to their mothers during the day and gradually become more independent. We occasionally catch females exiting a cave while carrying young, but most offspring are left behind when adults begin to forage.

Even though youngsters are left hanging in large groups, mothers are capable of recognizing and retrieving their own pups when they return. For example, a group of adult females and neonates was captured and banded during the day at a cave in Cuba (Silva-Taboada, 1979). After adults left to forage, marked pups were seen clinging to the rocks by themselves, but recapture of bats during the following day revealed that the mothers were carrying the same pups as when originally captured. How a brown flower bat locates and correctly identifies her offspring among the thousands of young bats that are in a cave is not yet known. However, other species, such as the Brazilian free-tailed bat, use a combination of spatial memory and auditory and olfactory cues to relocate their young (McCracken and Gustin, 1991), and the brown flower bat probably uses similar techniques.

At least five species of ectoparasite infest the brown flower bat on Puerto Rico (Gannon and Willig, 1994b, 1995). These include two streblid batflies (*Trichobius robynae* and *T. truncatus*), an argasid mite (*Ornithodoros viguerasi*), and spinturnicid wing mites (*Periglischrus cubanus* and *P. iheringi*). Quantitative data are sparse, but four of six bats captured in the Luquillo Experimental Forest carried ectoparasites, and up to 23 wing mites (*P. iheringi*) and 41 argasid ticks (*Ornithodoros* sp.) were recovered from individual bats (Gannon and Willig, 1995). An examination of a large number of these bats taken from their cave roosts likely would yield additional external parasite species. Other species of mite and batfly, as well as various tapeworms and roundworms, are reported from the Bahamas and Cuba (Silva-Taboada, 1979; Tamsitt and Fox, 1970a; Webb and Loomis, 1977).

Monophyllus redmani Leach, 1821
Greater Antillean Long-tongued Bat
Murciélago Lengüilargo

Taxonomy: *Monophyllus* combines with *Glossophaga* and 11 other genera to form the large phyllostomid subfamily Glossophaginae—a large group of nectar-feeding bats found throughout much of the New World (Simmons, 2005; Wetterer et al., 2000). Within the genus *Monophyllus* are two species, *M. redmani* and *M. plethodon,* and they are distinguished from each other by only slight differences in arrangement of the teeth. Today *M. plethodon* is restricted to the Lesser Antilles, but fossilized specimens of both species are known from caves on Puerto Rico. Presence of both bats in these cave deposits provides support for recognizing each as a distinct species, despite the seemingly minor difference in physical appearance (Choate and Birney, 1968; Schwartz and Jones, 1967). There are three subspecies of *M. redmani—M. r. clinedaphus, M. r. redmani,* and *M. r. portoricensis. M. r. portoricensis* is smaller than the other subspecies, and as its name suggests, it is the subspecies that occurs on Puerto Rico. The holotype of *M. redmani* is a specimen collected from Jamaica, and the type of *M. r. portoricensis* is a bat collected from a cave near Bayamón, Puerto Rico (Homan and Jones, 1975).

Name: Leach's (1821:73) original paper contained descriptions of seven genera with "foliaceous appendages" to the nose, and he considered some of the animals to have two nose-leaves (e.g., Old World *Nyctophilus).* The name *Monophyllus* comes from the Greek words for "one" and "leaf" and indicates the simpler structure of the nose-leaf in this bat. The specific epithet *redmani* apparently honors R. S. Redman, who gave the first specimen to Leach, whereas the subspecific designation refers to Puerto Rico, where this taxon is found. Wilson and Cole (2000) recently coined the name "Leach's single leaf bat" for the animal, but this name has not appeared in other literature.

Distribution and Status: The genus *Monophyllus* is endemic to the West Indies (Baker and Genoways, 1978). The Greater Antillean long-tongued bat is found on Cuba, Hispaniola, the southern Bahamas (*M. r. clinedaphus),* Jamaica (*M. r. redmani),* and Puerto Rico (*M. r. portoricencis;* Map 12). This species is common and broadly distributed on Puerto Rico.

Fig. 29 Portrait of a Greater Antillean long-tongued bat.

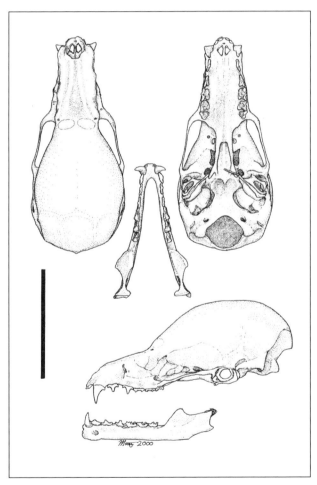

Fig. 30 Skull and mandible of a Greater Antillean long-tongued
bat (bar = 10 mm).

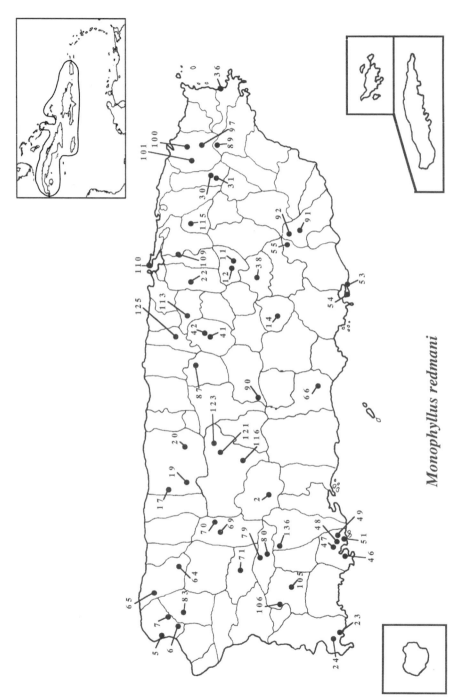

Map 12 Distribution of the Greater Antillean long-tongued bat on Puerto Rico and in the Caribbean basin. Location of numbered sites is detailed in Appendix 8.

Monophyllus redmani

Measurements and Dental Formula: Total length is 58–67 mm; length of tail, 8–10 mm; length of hind foot, 10–12 mm; height of ear, 11–14 mm; and length of forearm, 36–38 mm. Body mass is 6–10 g. Dental formula is incisors 2/2, canines 1/1, premolars 2/3, molars 3/3 = 34; however, one or more lower incisors may often be missing in adults (Phillips, 1971).

Description: The Greater Antillean long-tongued bat is the smallest phyllostomid on Puerto Rico. It has the unmistakable look of a nectar-feeding bat—elongate, narrow muzzle; long, highly protrusible tongue coated with bristlelike papillae; tiny lower incisors; and long, slender cheek teeth. The nose-leaf is short but well developed (Figs. 5, 29). As in other phyllostomids, the tail membrane is quite narrow but appears V-shaped when the legs are spread. As in the brown flower bat, the tail of the Greater Antillean long-tongued bat is short, only about half as long as the femur, and the posterior half of the tail extends freely beyond the border of the small interfemoral membrane. Flight membranes and ears are dark and naked. Dorsal fur ranges in color from smoky brown to gray, and ventral hairs typically are tipped with silver or cream, giving the venter a lighter-colored appearance. This species is most similar to the brown flower bat, but the long-tongued bat is distinguished easily by its smaller size and better-formed nose-leaf.

The long, narrow rostrum of this bat is quite distinctive (Fig. 30), although its skull is somewhat similar in shape to that of the brown flower bat. However, the skull of the long-tongued bat has a pinhole-sized foramen in the midline of the palate, anterior to the incisive foramina, that is lacking in the brown flower bat, although the skull must be well cleaned for this small opening to be seen. Another diagnostic feature is a broad gap between the right and left incisors in the mandible of the long-tongued bat that does not occur in a brown flower bat.

Natural History: The long-tongued bat forms large colonies containing up to a few hundred thousand individuals, and it is one of the most abundant species on Puerto Rico. During the day these bats, especially reproductive females, prefer to rest in hot caves, where they normally roost in association, with but spatially separated from, bats of other species (Rodríguez-Durán, 1998; Silva-Taboada, 1979). For example, an estimated 544,000 long-tongued bats lived in Cucaracha Cave prior to Hurricane Georges, along with 141,000 sooty mustached bats, and 43,000 ghost-faced bats (Jones et al., 2001; Rodríguez-Durán and Lewis, 1987).

Long-tongued bats seem to prefer roosting at temperatures near 33°C, but this phyllostomid probably can be found at a wider range of temperatures than any other Puerto Rican species of bat that roosts only in caves. Although long-tongued bats typically do not hang in contact with bats of other species, male long-tongued bats frequently form small clusters in cooler caves (Rodríguez-Durán, 1995). An entire cluster that we captured at Bonita Cave near Toa Alta consisted of 10 males, with one male Parnell's mustached bat in the cluster as well. Anthony (1918, 1925) noted clusters of long-tongued bats in Trujillo Alto Cave near Trujillo Alto on 14 July, and 80 of 83 bats that he examined were males. Even when clustered in cooler chambers, long-tongued bats are alert and readily fly when disturbed.

An extremely large number of bats concentrated in such a small area as a single cave means that the colony must disperse widely over the countryside each night to seek food. Not surprisingly, these bats appear to be strong fliers and frequently are captured in the Luquillo Mountains or are seen foraging in suburban San Juan, far from any known caves. This small species also is more adept at hovering than any other bat on Puerto Rico and can do so in spaces barely larger than its wing span. This ability certainly facilitates foraging among tree branches, but it also enables the long-tongued bat to escape from the harp traps commonly used by biologists to capture bats (Rodríguez-Durán and Lewis, 1987).

The Greater Antillean long-tongued bat is strictly nocturnal and begins to leave its roost after dark, generally between 28 and 69 minutes after sunset (Rodríguez-Durán and Lewis, 1987). Its onset of activity is earlier than in some species, such as the brown flower bat, but later than in others, such as the sooty mustached bat. The long-tongued bat often departs about the same time as the ghost-faced bat, but when both species roost in the same cave, they often use different routes to exit the cave.

The long-tongued bat is morphologically specialized for consumption of nectar, and 91 percent of 139 individuals sampled at Culebrones

Cave (Soto-Centeno, 2004) and 53 percent of 64 bats sampled at Cucaracha Cave (Rodríguez-Durán and Lewis, 1987) had at least some pollen on their fur or in their feces. In dry areas of southwestern Puerto Rico, long-tongued bats visit columnar cacti that bloom at night (Rivera-Marchand, 2001), and at other sites, swarms of long-tongued bats surround silk cotton trees that are in bloom. These bats are frequent visitors to flowers of banana and palm, even in the suburbs of San Juan, and pollen taken from bats at Culebrones Cave indicates that they also visit flowers of guava, maga, white siris, wild tamarind, and woman's tongue (Soto-Centeno, 2004). Following Hurricane Hugo in 1989, the number of long-tongued bats foraging in the Luquillo Mountains increased, presumably in response to the rapid increase in some flowering plants that occurred after the forest understory was opened to sunlight (Gannon and Willig, 1994a).

Greater Antillean long-tongued bats consistently consume insects and occasionally ingest fruit in addition to nectar. Soto-Centeno (2004), for example, found insect remains in fecal pellets from 73 percent of the bats that he sampled at Culebrones Cave. Compared with brown flower bats, however, long-tongued bats consume fewer hard-bodied prey, such as beetles, and more soft-bodied insects, especially flies and moths; long-tongued bats also take flying ants, true bugs, leafhoppers, and thrips, although the latter probably are ingested accidentally as the bats feed on nectar (Rodríguez-Durán and Lewis, 1987; Rodríguez-Durán et al., 1993; Soto-Centeno, 2004). The delicate teeth of the long-tongued bat are ill-suited for crushing fruit, and less than 25 percent of long-tongued bats produce fecal pellets containing seeds of any kind, which suggests that fruits are a minor component of the diet.

Data on reproduction are scarce, but births apparently occur at two different times during the year, as on Cuba (Silva-Taboada, 1979). On Puerto Rico, pregnant females are known from February through July and again in September and October. Records of lactating females, in contrast, currently exist only for April through July. Rodríguez-Durán (1984) reports an almost complete disappearance of adult males from a cave roost during March and May, which suggests the occurrence of maternity colonies at that time.

As in other phyllostomids, females give birth to only a single offspring. Neonates are naked with bright pink skin. An adult female carrying her young occasionally is captured in a harp trap outside a roost during evening emergence, but mothers usually leave the newborn at home during feeding bouts. Unlike brown flower bats, female long-tongued bats are not very attentive, and even during the day, we commonly observe pink neonates covering the walls of a maternity cave, roosting separate from their mothers and most other adults. Patterns of growth, mortality factors, and lifespan in this species are generally unknown. The only documented case of predation is a single individual found in the stomach of a Puerto Rican boa (Wiley, 2003).

Several species of mite and batfly infest the long-tongued bat on Puerto Rico (Gannon and Willig, 1994b, 1995). These include two spinturnicid wing mites (*Periglischrus iheringi* and *P. vargasi*), two spelaeorhynchid ear mites (*Spelaeorhynchus monophylli* and *S. praecursor*), and several streblid batflies (*Nycterophilia parnellii, Trichobius cernyi, T. intermedius, T. robynae,* and *T. truncatus*). The batfly *T. intermedius* is the most widespread parasite on long-tongued bats in the Luquillo Experimental Forest, occurring on 43 percent of the bats examined. Additional spinturnicid, macronyssid, and labidocarpid mites, streblid flies, and two roundworms are reported from Cuba (Silva-Taboada, 1979).

Stenoderma rufum Desmarest, 1820
Red Fig-eating Bat
Murciélago Frutero Nativo

Taxonomy: The genus *Stenoderma* is placed in the subfamily Stenodermatinae, along with *Artibeus* and 16 other genera (Simmons, 2005). However, unlike *Artibeus, Stenoderma* contains only a single species, *Stenoderma rufum,* which includes two living subspecies, *S. r. rufum* and *S. r. darioi.* The type specimen for *S. r. rufum,* and hence for the species, was taken from an unknown locality in the early 1800s, but given the present distribution of subspecies and the appearance of the type specimen, biologists believe that the original specimen was from the Virgin Islands. The description of *S. r. darioi* is based on a bat captured in 1965 near El Yunque Peak, in eastern Puerto Rico (Hall and Tamsitt, 1968).

Name: The word *Stenoderma* comes from the Greek words *stenos* and

Fig. 31 Portrait of a red fig-eating bat.

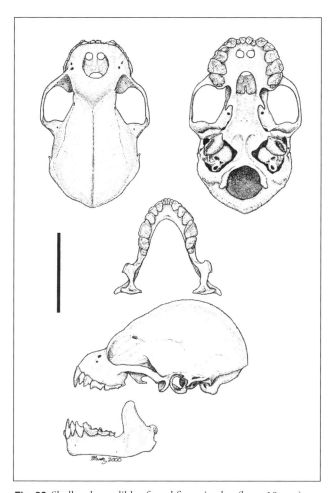

Fig. 32 Skull and mandible of a red fig-eating bat (bar = 10 mm).

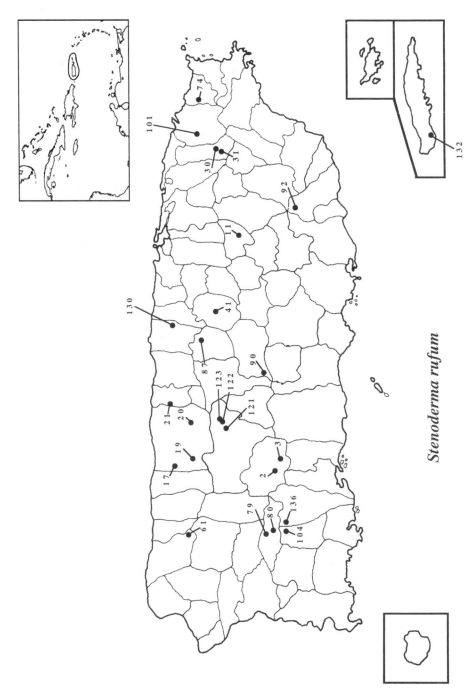

Map 13 Distribution of the red fig-eating bat on Puerto Rico and in the Caribbean basin. Location of numbered sites is detailed in Appendix 8.

Stenoderma rufum

derma, which mean "narrow" and "skin," respectively, most likely referring to the narrow tail membrane of this species. The specific epithet
rufum is Latin for red and refers to the occasionally reddish brown color
of this animal on the Virgin Islands, where the type presumably was
collected. The subspecific name *darioi* is in honor of Darío Valdivieso,
a neotropical biologist who published a number of papers concerning
bats in the 1960s and 1970s. Wilson and Cole (2000) suggest that the
common name "red fruit bat" is more appropriate for *S. rufum,*
although this alternative name has not been adopted yet by workers in
the field.

The history of the bat named *Stenoderma rufum* is intriguing and
shows the confusion that can result when museum specimens, particularly type specimens, are not cared for properly (Anthony, 1925).
Although A. Desmarest was the first to use the current scientific name
in 1820, this species actually had been introduced to the scientific
world a few years earlier as *le Sténoderme roux* by Étienne Geoffroy
Saint-Hilaire, a prolific French taxonomist of the early nineteenth century. Geoffroy Saint-Hilaire (1818) published a description of the
species based on the skin and skull of a specimen housed in the
National Museum of Natural History in Paris. However, the specimen
apparently had been mislabeled, and Geoffroy Saint-Hilaire reported
that it was from Africa. In 1869, W. Peters attempted to examine the
holotype, but he found only the skin; the skull apparently had disappeared since the time of Geoffroy Saint-Hilaire. A few years later, Peters
went to the same museum and examined a different specimen, this time
consisting of only a skull, which had been given the name *Artibeus unduatus* by another French taxonomist, P. Gervais. Peters noted that the
skull of *A. unduatus* appeared identical to the one that Geoffroy Saint-
Hilaire had drawn for his published description of *le Sténoderme roux.*
Peters surmised that the skull of *A. unduatus* was actually the long-lost
skull of *S. rufum,* and his conclusion generally is accepted today.

For almost a century, the red fig-eating bat was known only from
that single, poorly preserved specimen in the Paris museum. Over the
years, biologists had commented that the specimen's anatomical traits
strongly indicated that it represented a species from the New World and
not from Africa, but proof was lacking. Finally, in 1916, fossilized
remains of this species were found in caves on Puerto Rico, thus con-

firming its presence in the Western Hemisphere (Anthony, 1918). Nevertheless, whether the species was still living or had become extinct since its original description was not settled until 1957, when James Bee captured three individuals on St. John in the Virgin Islands (Hall and Bee, 1960).

Distribution and Status: The red fig-eating bat is endemic to the Greater Antilles and occurs on very few islands (Map 13). *S. r. rufum* is known from the Virgin Islands of St. John and St. Thomas, although it is quite rare there and we are not aware of any specimens taken on the Virgin Islands in the last 30 years. *S. r. darioi* was thought to live only on Puerto Rico, but a juvenile red fig-eating bat was captured recently on Vieques, suggesting the presence of an additional population on that island (Rodríguez-Durán, 2002). This species appears most common in the Luquillo Experimental Forest, where at one time it represented at least 25 percent of bats captured in tabonuco forest (Willig and Bauman, 1984). Only about 50 individuals have been taken in other parts of Puerto Rico and on Vieques (Gannon, 2002), but whether the paucity of records away from Luquillo represents a habitat preference by the species or simply lack of fieldwork in habitats other than tabonuco forest is unknown.

The International Union for Conservation of Nature and Natural Resources classifies the red fig-eating bat as vulnerable to extinction (category "VU a1c"; Hutson et al., 2001). This designation indicates that although not critically endangered at present, the species may be facing a high risk of extinction in the medium-term future. The red fig-eating bat has the smallest geographic distribution of any species of bat that occurs on Puerto Rico and one of the smallest in the Antilles. The IUCN declared the species vulnerable because of a suspected decrease in the already small geographic distribution of this island endemic and because of the devastating effects of hurricanes Hugo and Georges (see Natural History).

Measurements and Dental Formula: Total length is 60–69 mm; length of hind foot, 12–18 mm; height of ear, 15–20 mm; and length of forearm, 48–52 mm. Body mass is 20–31 g. As in many species of bat, female red fig-eating bats are marginally but consistently larger than males in many traits (Gannon et al., 1992; Jones et al., 1971). Dental formula is incisors 2/2, canines 1/1, premolars 2/2, molars 3/3 = 32.

Aerial photograph of haystack hills (*mogotes* or *pepinos*) in the northeastern karst area of Puerto Rico. PHOTO BY J. COLÓN.

Hurricanes initiate numerous landslides, such as this one in tabonuco forest near El Verde. These catastrophes alter the landscape substantially, essentially returning portions of the ecosystem to stages without appreciable life in or on the soil. PHOTO BY G. R. CAMILO.

Fig.3a

Fig.3b

Hurricane-mediated disturbances are common in Puerto Rico as evinced by the contrast between images of this forested site (a) immediately before and (b) after Hurricane Hugo.
PHOTOS BY B. HAINES.

Fig.**4a**

Fig.**4b**

Two extreme habitats found on Puerto Rico. The upper photo shows dry coastal forest (subtropical dry forest of Holdridge) near Guánica, on the extreme southwestern edge of the island, where a rain shadow occurs. The lower photo illustrates upper Luquillo forest (lower montane wet forest of Holdridge) on El Yunque, in the northeastern portion of the island, near Luquillo.

UPPER PHOTO BY J. COLÓN; LOWER PHOTO BY M. R. GANNON.

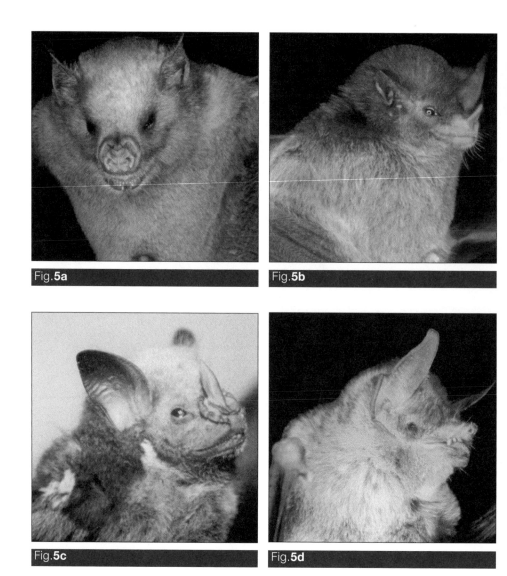

Four species of Puerto Rican bats endemic to the Caribbean region: (a) Antillean fruit bat, (b) Greater Antillean long-tongued bat, (c) red fig-eating bat, and (d) sooty mustached bat.
PHOTOS BY M. R. GANNON.

Fig.11

A group of Jamaican fruit bats roosting in a solution cavity in the ceiling of a cave in Ciales.
PHOTO BY J. COLÓN.

Plants that grew from seeds eliminated in the feces of Jamaican fruit bats roosting during the day in a cave called Cueva de Las Larvas, south of Arecibo. Although these plants soon died from lack of sunlight, similar seeds defecated at night, while bats are flying or feeding in trees, help sustain populations of native plants. On Puerto Rico, such underground collections of dead seedlings are termed *jardines pálidos,* "pallid gardens," because of their pale color. Note also fragments of green leaves in the foreground; frugivorous bats occasionally supplement their diet with leaves. PHOTO BY M. KURTA.

Pre-Columbian artifacts from Puerto Rico resembling bats: (a) leaf-nosed bat, probably representing a Jamaican fruit bat; (b) face of a bat, possibly a velvety free-tailed bat: (c) bat with elongated snout; and (d) a bat face with a short snout and lacking a nose-leaf, possibly modeled after the Antillean fruit bat. Specimens are housed in the Museum of the Municipality of Manatí.

PHOTOS BY A. RODRÍGUEZ-DURÁN.

Fig.28

A Puerto Rican boa feeding on a brown flower bat at the entrance to Culebrones Cave, south of Arecibo. PHOTO BY M. R. GANNON.

Description: Although red fig-eating bats from the Virgin Islands may have a reddish tint to their fur, those from Puerto Rico typically appear brown or tan. Ventral hairs are tipped with gray and appear somewhat lighter. A dab of white occurs on the side of the head below each ear, and both sexes also have a white shoulder patch that is about 4 millimeters in diameter (Figs. 5, 31). Underneath the shoulder patches of adult males are glands that emit a strong musklike odor. A tail is absent, and the narrow interfemoral membrane has only a sparse covering of hairs. A well-developed nose-leaf easily establishes this species as a member of the Phyllostomidae, and the bat's gentle disposition combined with medium body size, white patches of fur, and lack of a tail distinguish this animal from all other phyllostomids on Puerto Rico.

Lack of W-shaped ridges on the cheek teeth and presence of incisive foramina in the palate indicate that the skull belongs to one of the phyllostomids of Puerto Rico (Fig. 32). The rostrum of a red fig-eating bat differs from that of other phyllostomids on the island in that it is obviously wider than it is long. Skulls of the Jamaican and Antillean fruit bats are moderately similar in construction, but their rostrums are not as short and the greatest length of their skulls (more than 26 mm) exceeds that of a red fig-eating bat (21–24 mm).

Natural History: Only the population in the Luquillo Mountains has been studied extensively, and almost all knowledge of the natural history of this phyllostomid comes from animals living there. The red fig-eating bat, like other members of its subfamily, is primarily a frugivore, but despite its generally accepted common name, there is no evidence that it actually eats figs. In the Luquillo Mountains, the most commonly eaten fruits are from the trumpet tree, bullet-wood, and sierra palm, with a few bats taking elder as well (Genoways and Baker, 1972; Scogin, 1982; Willig and Bauman, 1984; Willig and Gannon, 1996). Frugivorous bats play an important role as seed dispersers in tropical ecosystems, and the red fig-eating bat may be the major—and perhaps only—disperser of bullet-wood seeds in the tabonuco forest of Puerto Rico (You, 1991). Foraging occurs both above and below the forest canopy (Jones et al., 1971).

Daily movements of the red fig-eating bat have been documented extensively in tabonuco forest by radio tracking (Gannon, 1991; Gannon and Willig, 1994a). Unlike other phyllostomids that live in large colonies in caves, both male and female red fig-eating bats are solitary and roost among the leaves of the forest canopy. These bats frequently change their roosting location, and sites are seldom occupied more than once.

Home range is small, only about 2.5 hectares on average. Subadults have larger home ranges than do mature adults, perhaps because young animals are less experienced at finding food or possibly because adults exclude them from trees with the best quality or greatest quantity of fruit. In any event, there is no evidence of territoriality, and individual home ranges overlap considerably. The small home range indicates that the red fig-eating bat spends little time or energy commuting to feeding areas. This behavior contrasts markedly with that of other fruit-eating species, such as the Jamaican fruit bat, which roost in colonies and often travel great distances to foraging grounds (Morrison, 1978a, 1978b).

Red fig-eating bats are active throughout the night and begin roosting just before dawn. Although some species of bat avoid flying in bright moonlight, presumably to avoid predation (Morrison, 1978c), the red fig-eating bat, like the Jamaican fruit-eating bat, does not alter its behavior in response to changing lunar illumination on Puerto Rico (Gannon and Willig, 1997). This lack of a response to moonlight may be related to the fact that the red fig-eating bat evolved on isolated islands with no visually oriented nocturnal predators capable of capturing this phyllostomid.

Hurricane Hugo passed over the Luquillo Experimental Forest in September 1989 and caused large-scale damage and defoliation (Walker, 1991; Walker et al., 1996). Immediately after the hurricane, captures of the red fig-eating bat decreased by 80 percent, and population levels did not return to pre-hurricane levels until five years following the storm (Gannon, 2002). Although juveniles made up 30–40 percent of the catch before Hurricane Hugo, this proportion fell to 10–20 percent after the storm, demonstrating an effect on reproductive success as well. This foliage-roosting species was dealt another severe blow by Hurricane Georges in 1998, and the population in the Luquillo Experimental Forest showed no sign of recovery over four years later (Gannon, 2002).

Hurricane Hugo greatly reduced availability of fruits eaten by this species, and bats that remained in the forest responded by increasing size of their home range by a factor of five. Even though number of red fig-eating bats greatly declined, this species was still the most abundant frugivorous bat in tabonuco forest during the first year after the hurricane (Gannon and Willig, 1994b). Consequently, the red fig-eating bat probably played an important role in disseminating seeds of early successional plants following the storm.

Pregnant females have been captured on Puerto Rico in January, March, June, July, and August, and lactating bats are known from March, May, June, and July. Males with descended testes have been taken in January, February, March, June, July, and August. As with many species of bat on Puerto Rico, no data on the reproduction of red fig-eating bats are available for September through December.

Females often are simultaneously pregnant and lactating, which indicates that females most likely undergo a postpartum estrus and that the species is polyestrous (Gannon and Willig, 1992; Genoways and Baker, 1972; Willig and Bauman, 1984). Nonetheless, births do not appear to be restricted to particular times of year; that is, red fig-eating bats exhibit asynchronous polyestry. This reproductive pattern is characteristic of species that live in areas where the abundance of food does not vary greatly on a seasonal basis (Wilson, 1973, 1979), and this appears true in the tabonuco forest of the Luquillo Mountains (Devoe, 1990).

Among mammals, bats have the largest offspring relative to the size of the mother, and among bats, phyllostomids have some of the largest neonates (Barclay and Harder, 2003; Hayssen and Kunz, 1996; Kurta and Kunz, 1987). In the red fig-eating bat, the single newborn weighs about 37 percent of the mother's body mass after she has given birth, and its forearm already is 59 percent as long as that of its mother (Tamsitt and Valdivieso, 1966). Young of this species are born with dense grayish hairs covering the back, and the white shoulder patch is obvious at birth. The face, in contrast, is mostly naked and appears pink. Eyes are open at birth. Females give birth in the usual resting position with the head facing down; youngsters are delivered headfirst and received by the mother in cupped wings.

The only ectoparasites reported for the red fig-eating bat are four species of mite (Gannon and Willig, 1994b, 1995), including one spinturnicid (*Periglischrus iheringi*) and three labidocarpids (*Paralabidocarpus artibei, P. foxi,* and *P. stenodermi*). The wing mite *P. iheringi* is the most common ectoparasite and occurs on 35 percent of red fig-eating bats, with infested individuals harboring about four mites each on average. As in the Jamaican fruit bat, juveniles support a greater number of parasites than do adults, perhaps because adults are more efficient at grooming and removing the mites.

Levels of infestation are low compared with other Puerto Rican species of bat that have been studied, and this may be related to the solitary roosting habit of the red fig-eating bat. Roosting alone would limit opportunities to obtain parasites by direct contact with other individuals, and this solitary behavior also would minimize the possibility of uninfected animals directly contacting parasites like argasid ticks, which spend part of their life cycle off the host and in the roost area.

FAMILY VESPERTILIONIDAE
Plain-nosed or Vesper Bats

With at least 6 subfamilies, 49 genera, and over 325 species, the Vespertilionidae is the second largest family of mammals and the largest family of bats (Hoofer and Van Den Bussche, 2003; Simmons, 2005). Vespertilionids have a cosmopolitan distribution, and in the New World, the family ranges from above the Arctic Circle to the tip of South America. A number of species have invaded the Antilles, and strong fliers have colonized even some of the most remote oceanic islands. The hoary bat, for instance, is the only native land mammal on the Hawaiian Islands. Two of the 13 species of Puerto Rican bats belong to this family.

Despite their taxonomic diversity, vespertilionids generally are small bodied, ranging in size from 2 or 3 grams in the lesser bamboo bat of Southeast Asia to perhaps 75 grams in the great noctule of Europe and the great evening bat of Asia (Nowak, 1999). A nose-leaf is absent in "plain-nosed" bats, and the eyes always appear small and beady. Ears vary from short, rounded pinnae that barely reach the top of the skull, as in the red bat, to spectacular appendages that equal the body in length, as in the spotted bat. The tail membrane is well developed, and the tail vertebrae usually extend to the membrane's posterior edge and sometimes slightly beyond (Fig. 9). Most vespertilionids are various shades of brown, black, or gray, but some appear red, yellow, or orange. A few have spots, and some, such as the silver-haired bat, have many hairs tipped with white, which results in an overall "frosted" appearance.

Most species are cave dwellers, but many use other natural shelters including rocky crevices, hollow trees, loose bark, tree stumps, and foliage (Nowak, 1999). The two species of bamboo bats have suction cups on their thumbs and feet that allow these animals to roost inside smooth-walled stalks of bamboo. Other vespertilionids readily occupy manmade structures that resemble caves or rock crevices, including mines, tunnels, sewers, culverts, old wells, and concrete bridges. In addition, many species roost in buildings.

Like roosting behavior, social behavior is highly varied—as one might expect in a group containing so many species (Nowak, 1999). Some vespertilionids roost alone, some are found in pairs or small groups, and still others form colonies containing hundreds or thousands of individuals. In many vespertilionids, females are colonial during pregnancy and lactation, while males lead a solitary existence.

Mating strategies are not known for most species, but a number of vespertilionids form harems, with a single male actively defending a group of 5–20 females or defending instead a resource that attracts a group of females. A male common pipistrelle from Europe, for example, excludes other adult males from a tree hollow or similar structure that females need for roosting, thereby controlling access to the females when they come into heat (Gerell and Lundberg, 1985). Mating by the little brown bat of North America, in contrast, has been described as "random and promiscuous" (Wai-Ping and Fenton, 1988:643), whereas the wooly bat of Southeast Asia is monogamous (McCracken and Wilkinson, 2000).

Vespertilionids are primarily insectivorous. Flying insects typically are scooped out of the air using a wing or tail membrane and passed to the mouth, often while the bat continues flying. The pallid bat (*Antrozous pallidus*) and a number of other long-eared species, in contrast, are gleaners, capable of plucking at least some of their prey from the surface of the ground or a tree trunk. At least one vespertilionid, the Mexican fish-eating bat, has strayed from a life of insectivory and consistently feeds on small fish, making it the ecological equivalent of Puerto Rico's greater bulldog bat (Stadelmann et al., 2004). Whatever the prey, vespertilionids rely on echolocation to find their food; each bat produces high-frequency sounds in the larynx, emits them through the mouth, and listens for the returning echoes.

Eptesicus fuscus (Palisot de Beauvois, 1796)
Big Brown Bat
Murciélago Ali-oscuro

Taxonomy: *Eptesicus fuscus* is one of 23 species within the genus *Eptesicus,* a widespread taxon that is found on all continents except Antarctica (Simmons, 2005). There are 11 subspecies of the big brown bat, and the one living on Puerto Rico is called *E. f. wetmorei* (Kurta and Baker, 1990). The original description of *Eptesicus fuscus* is based on a specimen captured in Philadelphia, Pennsylvania, in the late eighteenth century, whereas the type for the subspecies *E. f. wetmorei* was collected in 1912 in Maricao, Puerto Rico.

Name: *Eptesicus* means "house flyer," and this bat is so named because in North America, it commonly roosts in human dwellings (Rafinesque, 1820). The specific designation *fuscus* refers to the rather dark or dusky color of its fur, and the subspecific epithet *wetmorei* is in honor of Alex Wetmore, who collected the first specimens of the subspecies (Jackson, 1916). The common name most likely refers to the fact that this bat is larger than the little brown bat, which is another North American species that typically occupies houses and barns.

Distribution and Status: The big brown bat has an extensive distribution in the New World (Map 14). This bat inhabits southern Canada and virtually all the continental United States. Its range continues through Mexico and Central America into northern South America, where it is found in northern Colombia and western Venezuela. Within the Caribbean basin it occurs on all islands of the Greater Antilles and on Barbados and Dominica in the Lesser Antilles (Kurta and Baker, 1990; Simmons, 2005). The large distance between Dominica and Barbados and the location of other known populations suggest that systematic sampling of bats on additional islands of the Lesser Antilles may yield new distributional records for the big brown bat.

Anthony (1925) noted that this mammal was uncommon on Puerto Rico, and that is our impression as well. Its seeming rarity, however,

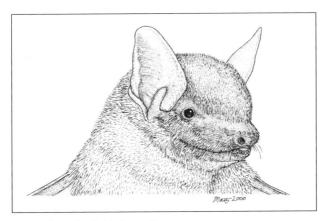

Fig. 33 Portrait of a big brown bat.

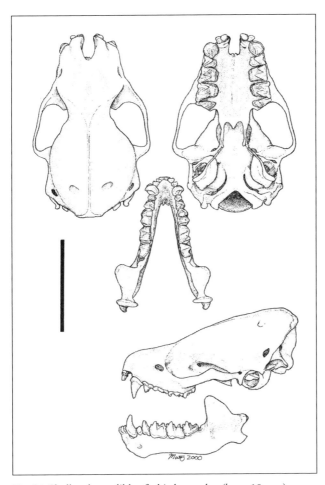

Fig. 34 Skull and mandible of a big brown bat (bar = 10 mm).

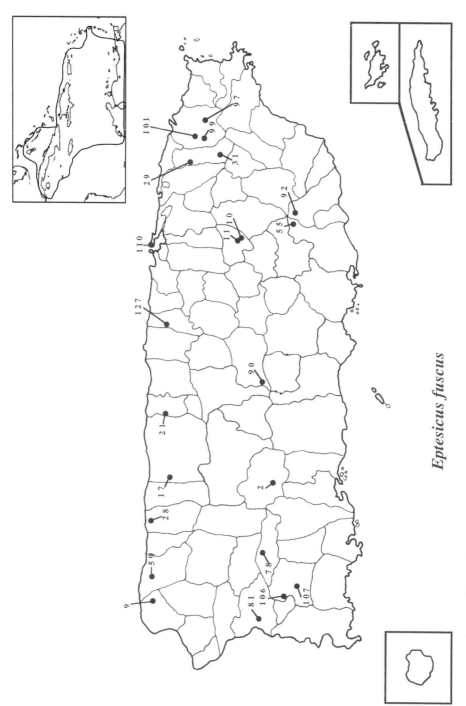

Map 14 Distribution of the big brown bat on Puerto Rico and in the Caribbean basin. Location of numbered sites is detailed in Appendix 8.

Eptesicus fuscus

may be related partly to the difficulty of capturing this species using ground-based mist nets, which are typically only 2.5 meters in height. The big brown bat usually forages above 4 meters (Kurta, 1982) and often flies over the forest canopy. In any event, this vespertilionid is captured most frequently in forested sites, such as in the Luquillo Mountains or along the Cordillera Central, and seems rarer along the coast. A pattern of greater abundance in forested highlands also is seen in populations of this species living in tropical areas of Mexico and Central America (Kurta and Baker, 1990), but fieldwork directed at this species is needed to determine more completely its distribution on Puerto Rico.

Measurements and Dental Formula: Total length is 109–126 mm; length of tail, 35–57 mm; length of hind foot, 8–14 mm; height of ear, 14–20 mm; and length of forearm, 45–51 mm. Body mass is 13–21 g. Dental formula is incisors 2/3, canines 1/1, premolars 1/2, molars 3/3 = 32.

Description: The big brown bat has a broad head and nose and a robust body (Fig. 33). Ears are thick, short, and somewhat rounded at the top. The plain tragus is broad at its base but narrows and turns slightly forward near the tip. The uropatagium is well developed. Unlike the tail vertebrae of a mormoopid, those of a big brown bat extend to the posterior margin of the tail membrane and are encased by the uropatagium along their entire length. The face, ears, wings, and tail membrane are essentially naked and dark brown to black in color, whereas body fur is chestnut to chocolate brown above and paler brown below. Overall body size, structure of the tail and uropatagium, and the distinctive brown color make it easy to separate this species from others on Puerto Rico.

Skulls of vespertilionid bats, such as the red bat and big brown bat, have a distinct U-shaped notch in the front of the palate that is widest at the front of the notch (Fig. 34). Brazilian free-tailed bats have a similar indentation of the palate, but the widest point of their notch occurs well behind the opening. Big brown bats and red bats are separated easily from each other by size; greatest length of the skull is 19–20 millimeters in big brown bats but only 13 millimeters in red bats. In addition, red bats have one pair of upper incisors, whereas the big brown bat has two pairs.

Natural History: The big brown bat is one of the most studied species of bat in the United States and Canada, but comparatively little is known about its biology in tropical regions (Kurta and Baker, 1990). In North America it frequently roosts in buildings, and especially in the West, it uses hollow trees. Nevertheless, over 75 percent of daytime retreats on Cuba are in caves, and the remainder are in manmade structures (Silva-Taboada, 1979). Individuals or small groups commonly roost in the entrances to these caves, where it is cool and often quite bright, and the bats usually tuck themselves into fissures or solution cavities rather than hang in the open. Larger groups, up to a few hundred animals, may form tight clusters on a wall or ceiling in deeper portions of a cool cave. Big brown bats usually are the only species in these clusters, but occasionally a Brazilian free-tailed bat may join. When roosting in buildings, big brown bats on Cuba most often seek shelter under roofing tiles.

Anthony (1925:58) reported big brown bats on Puerto Rico from caves near San Germán and Pueblo Viejo, where the bats were "using deep crevices and holes in the limestone in much the same fashion as did *Tadarida,* being crowded into its retreats in large numbers." Similarly, Schwartz (in Timm and Genoways, 2003) captured a few big brown bats in shallow caves on the western slopes of the Cordillera Central during the mid-1900s. Today the few documented dayroosting sites on Puerto Rico are also in caves or cavelike structures, such as abandoned tunnels and culverts under roads. One culvert near Aguadilla contained at least 22 big brown bats, although the bats typically roosted alone or with only one or two other individuals.

Compared with other bat species on Puerto Rico, big brown bats are quiet animals, producing few audible vocalizations, even when disturbed (Silva-Taboada, 1979). On Cuba they begin leaving the roost near sunset, from about 10 minutes before to 19 minutes after sunset, and return between 25 and 7 minutes before sunrise. Foraging occurs throughout the night but is punctuated by intermittent stops at a nightroost for rest or perhaps to dismember large prey. We have discovered this species nightroosting under a highway bridge over the Río Guajataca and in a building at Hacienda Buena Vista, near Ponce. On a warm night in July, about two hours after sunset, we observed 34 big brown bats nightroosting at the hacienda; two animals roosted by themselves, but the others clustered in groups of 3, 7, 10, and 12 bats. These clusters were very tight, even though air temperature in the building likely was 26°C or greater. On another occasion we witnessed six big brown bats clustered with a single Brazilian free-tailed bat at the hacienda.

The big brown bat is a habitat generalist, hunting in a variety of situations, including over water, at forest edges, and in open- and closed-canopy forests that are not too cluttered (Kurta and Baker, 1990). This bat is an aerial insectivore that preferentially feeds on beetles throughout its range, although a variety of other insects supplements the diet (Kurta and Baker, 1990). For example, fecal pellets from a single female from Puerto Rico contained the remains of beetles and true bugs (Rodríguez-Durán et al., 1993), and 72 percent of the stomachs from Cuban big brown bats contained beetles, although moths, flies, ants, roaches, and/or lacewings were present in 4–12 percent. Body mass may increase by 2.9 grams, or 24 percent, after a single foraging bout (Silva-Taboada, 1979), and in late lactation, females typically consume their own weight in insects each night (Kurta et al., 1990). Insects are detected using echolocation calls that are the lowest in frequency of any bat on Puerto Rico, typically sweeping from 40 or 50 kilohertz to about 25 kilohertz (Fenton and Bell, 1981; Simmons et al., 1979). Flight speed in the open is 33 kilometers per hour.

On Cuba, one to four embryos are present in the uterus very early in pregnancy, although a maximum of two young are carried to term (Silva-Taboada, 1979). Twins are born more than 85 percent of the time. Pregnant females are found as early as March on Cuba, but the earliest known date of birth on that island is 16 June. Anthony (1925:58), in contrast, reports the presence of young "half the size of the parent" on Puerto Rico on 6 June. In temperate areas, big brown bats mate in autumn, but females store the sperm in the uterus throughout hibernation and do not ovulate until spring (delayed fertilization). Whether autumn mating and sperm storage occur in tropical, nonhibernating populations is unknown.

While giving birth, the big brown bat, like most vespertilionids, roosts head up and uses its broad interfemoral membrane as a basket to catch the young as it leaves the birth canal. Breech birth is the norm. A big brown bat is born without fur and with eyes closed and ears folded.

In litters containing only one offspring, the single young weighs 26 percent of the mother's body mass, whereas the combined mass of twin fetuses represents 30 percent of the mother's mass (Silva-Taboada, 1979). On the mainland, most young begin to fly in three to four weeks and consume at least some milk for up to five weeks after birth (Burnett and Kunz, 1982; Kurta et al., 1990). Maximum longevity for a big brown bat is 19 years in the wild and almost 22 years in captivity (French, 2002; Kurta and Baker, 1990).

Big brown bats in temperate areas consistently use torpor, which is a short-term lowering of body temperature. This is an energy-saving mechanism that bats use when cold temperatures or prolonged rains make it difficult to obtain sufficient energy in the form of flying insects. Apparently big brown bats from tropical areas have this physiological ability as well. Silva-Taboada (1979), for example, mentions two females in a dayroost that had body temperatures of only 26–27.5°C, when air temperature was 19°C. The two bats were able to raise their body temperature to 38°C in 20 minutes. Although big brown bats from temperate areas continue to forage at ambient temperatures down to 10°C, those in Cuba do not even leave their roost when air temperatures fall below about 20°C, preferring to remain torpid for the evening.

Temperate vespertilionids, such as the big brown bat, also hibernate. Hibernation is similar to torpor in a physiological sense, but the bats maintain the lowered body temperature throughout winter, except for occasional arousals to drink and urinate. Hibernation evolved in response to the seasonal absence of prey (flying insects) in cold climates, and not surprisingly, there is no evidence for hibernation in the tropics, where insects are available year-round. However, temperate bats store fat in autumn for later use as metabolic fuel during hibernation, and tropical populations still display a seasonal cycle in body mass. Big brown bats on Cuba, for example, show a marked increase in body mass in autumn and then lose an average of 32 percent of their body mass from November to March (Silva-Taboada, 1979). This time of mass loss coincides with the coldest conditions of the year on Cuba and the greatest use of daily torpor by the bats.

Although endoparasites of the big brown bat have not been examined on Puerto Rico, they have been studied extensively on Cuba

(Rutkowska, 1980; Zdzitowiecki and Rutkowska, 1980). Endoparasites from Cuban big brown bats include flukes, tapeworms, and roundworms, and one would assume that Puerto Rican individuals harbor similar parasites. Only two ectoparasites are known from Puerto Rico, and both are spinturnicid mites (*Periglischrus iheringi* and *Spinturnix bakeri*) that were taken from big brown bats captured in the Luquillo Experimental Forest (Gannon and Willig, 1994b, 1995; Willig and Gannon, 1996). Streblid flies, an argasid tick, and macronyssid, spinturnicid, labidocarpid, and sarcoptid mites are documented ectoparasites on Cuba. Many other internal and external parasites have been identified in big brown bats taken from continental populations (Kurta and Baker, 1990).

Lasiurus borealis (Müller, 1776)
Red Bat
Murciélago Rabi-peludo

Taxonomy: At one time, biologists believed that the red bat was a single species that occurred throughout most of the New World, including the Greater Antilles and the Bahamas (Koopman, 1993). Recent biochemical and chromosomal evidence, however, indicates that populations in eastern North America (*Lasiurus borealis*) probably are distinct from those in western North America and in South America (*Lasiurus blossevilli*; Morales and Bickham, 1995). In addition, a few molecular systematists suggest that some Antillean populations that traditionally have been called red bats and assigned to *L. borealis* may be related more closely to Seminole bats (*Lasiurus seminolus*), and perhaps these island residents should be considered subspecies of that group rather than of the red bat (Morales and Bickham, 1995). Simmons (2005), in contrast, proposes that these island populations are not subspecies but three distinct species: *L. pfeifferi* (Cuba), *L. degelidus* (Jamaica), and *L. minor* (Puerto Rico, Hispaniola, and Bahamas).

Given this uncertainty, we follow Koopman (1993) for the time being and retain the red bat, *L. borealis,* as a single widely distributed species and *L. b. minor* as the subspecies found on Puerto Rico. The type of *L. borealis* is a specimen collected in New York near the time of the American Revolutionary War, whereas the original description of

Fig. 35 Portrait of a red bat.

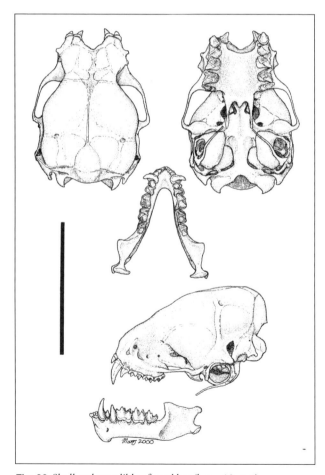

Fig. 36 Skull and mandible of a red bat (bar = 10 mm).

L. b. minor is based on a skull that was discovered in an owl pellet in a Haitian cave over 150 years later (Miller, 1931; Shump and Shump, 1982).

Name: *Lasiurus* comes from the Greek words *lasios,* meaning shaggy or hairy, and *oura,* meaning tail (Jaeger, 1955). The specific epithet *borealis* is Latin and refers to the north, presumably a reference to the part of the United States where the type of this species was obtained. The subspecific name *minor* denotes the small size of these red bats relative to those living on Cuba and Jamaica (Miller, 1931). The generic name *Nycteris* sometimes is applied incorrectly to the red bat and other lasiurines (e.g., Hall, 1981).

Distribution and Status: The red bat occurs throughout much of eastern North America, ranging from southern Canada to Texas and Florida. In addition, this mammal occurs on all the Greater Antilles as well as the Bahamas (Map 15; Koopman, 1993). The subspecies *L. b. minor* is restricted to Hispaniola, Puerto Rico, and the Bahamas.

Red bats were not known from Puerto Rico until May 1961 (Starrett and Rolle, 1962), when a male collided with the windshield of a moving car near the town of Moca. Since that time, only six other individuals have been captured, and all were taken during June or July. Three individuals were caught in the Luquillo Experimental Forest, two were netted in the Susúa State Forest, and one was found on the campus of the University of Puerto Rico in Río Piedras. The red bat is a solitary animal that frequently flies high above the forest canopy, and consequently its abundance on Puerto Rico, like that of the big brown bat, may be underestimated by traditional surveys that emphasize netting in the understory.

Measurements and Dental Formula: Total length is 98–104 mm; length of tail, 40–47 mm; length of hind foot, 7–10 mm; height of ear, 13–15 mm; and length of forearm, ca. 42 mm. Body mass is 6–12 g. Dental formula is incisors 1/3, canines 1/1, premolars 2/2, molars 3/3 = 32.

Description: The red bat is a very handsome mammal, with short, rounded ears and a compact rostrum (Fig. 35). The long, narrow wings are largely naked, although a patch of white hairs usually occurs at the base of the thumb and along the proximal part of the second metacarpal (the elongated digit next to the thumb). The tail reaches to the posterior edge of the tail membrane, and unlike that of all other bats on Puerto

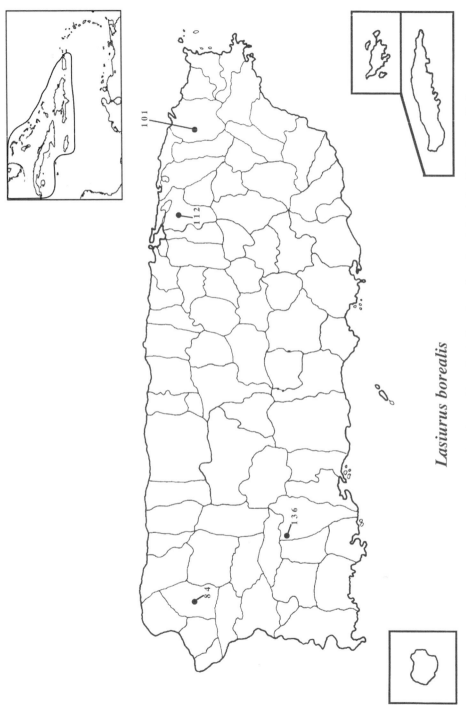

Map 15 Distribution of the red bat on Puerto Rico and in the Caribbean basin. Location of numbered sites is detailed in Appendix 8.

Lasiurus borealis

Rico, the tail membrane of the red bat is heavily furred. This species also is distinguished easily from other bats by its distinctive color. The dorsal surface varies from a rusty reddish orange to yellowish red, and many hairs are tipped with white, giving the animal a frosted appearance. Adult males generally have less "frosting" and often appear more brightly colored than adult females. Such sexual dimorphism is lacking among immature males and females when they first start to fly, and both resemble adult females at that time. In all red bats, the deep black of the wing membranes contrasts markedly with the reddish fur. Female red bats differ from all other species of bat on Puerto Rico in having four rather than two nipples.

Red bats and big brown bats, the two vespertilionids on Puerto Rico, have a broad, U-shaped notch in the palate that is not found in any other species on the island (Fig. 36). Red bats and big brown bats are distinguished easily from each other by number of incisors on each side of the notch—one in the red bat and two in the big brown bat. Although the Brazilian free-tailed bat also has a single upper incisor on each side and a notch in its palate, the indentation in that species is widest toward the rear of the notch rather than near the opening, as in vespertilionids.

Natural History: In temperate North America red bats roost alone, or at most, a mother will hang near her dependent offspring. These bats are similar to red fig-eating bats in that they rest among the leaves of a tree and do not take shelter in tree hollows, buildings, or caves. No roost trees have been found yet on Puerto Rico, but on the mainland, red bats use various species of tree, including hickories, oaks, sweetgum, and tulip tree (Hutchinson and Lacki, 2000; Mager and Nelson, 2001; Menzel et al., 1998). Individual bats are not loyal to any particular roost but change trees about once every day or two; the new roost is usually nearby and always within a few hundred meters of the previous one.

Red bats are excitable animals. They make a distinctive, audible sound that combines hisses and clicks whenever they are caught in a net, handled, or otherwise disturbed. When confined to a collecting bag or cage, a red bat commonly thrashes about, often bruising its wrist or other parts of the wing skeleton.

These bats are swift fliers but not highly maneuverable; consequently, they typically forage in open areas—above the canopy, in woodland openings, or along forest edges. In areas inhabited by humans, red bats often feed on insects attracted to street lights or isolated lights on buildings (Hickey and Fenton, 1990). Throughout North America, red bats feed primarily on moths, but they also consume varying amounts of other prey, such as beetles, leafhoppers, flies, and ants. Diet in tropical areas is largely unstudied. On Cuba only a few fecal pellets from a single bat have been examined, and they contained only beetles (Silva-Taboada, 1979). On Puerto Rico all nine fecal pellets produced by one female red bat contained moths, and two of the nine pellets also contained winged termites (Rodríguez-Durán, 1999). The stomach contents of a different red bat from the Luquillo Mountains included a large number of flying ants (Willig and Gannon, 1996).

As do other vespertilionids, bats in the genus *Lasiurus* detect obstacles and flying prey by using frequency-modulated echolocation calls that are emitted through an open mouth. However, calls of lasiurines differ from those of most other species in that the minimum frequency of individual pulses within a sequence is not constant but fluctuates unpredictably over a few kilohertz (O'Farrell et al., 1999). For example, minimum frequency of pulses produced by red bats generally is near 40 kilohertz, but minimum frequency within a sequence can vary from 36 to 46 kilohertz. This distinctive variation allows biologists with ultrasonic detectors to distinguish flying red bats from other Puerto Rican species, and this new technique likely will play a major role in future efforts to determine the distribution of this uncommon species on the island.

Red bats from temperate areas frequently undergo daily torpor in summer in response to food shortages. All apparently hibernate in winter, and those living in Canada and the northern United States migrate south in autumn to milder climates, where they eventually hibernate in trees or forest litter (Moorman et al., 1999). The presence of red bats, especially reproductive females, on Puerto Rico in late spring and summer suggests that red bats are year-round residents of the island and not just winter migrants from cold climates. Although Puerto Rico is not cold enough to elicit hibernation by this vespertilionid or any other species, red bats on Puerto Rico probably resort to torpor during cold snaps, in a manner similar to the big brown bat on Cuba (Silva-Taboada, 1979).

Worldwide, perhaps 90 percent of all bat species produce a single young in each litter, and most of the rest have twins. Bats in the genus *Lasiurus,* however, are unique among bats in the New World in having four functional mammary glands, and consequently, they are capable of consistently supporting litters containing more than two young. Litter size in red bats varies from one to five on the mainland, with an average of 2.3 (Shump and Shump, 1982). A litter of five red bats from Louisiana is the largest litter ever reported for any species of bat (Hamilton and Stalling, 1972).

In temperate areas red bats typically give birth to naked young in June, and lactation lasts five to six weeks (Shump and Shump, 1982). Reproductive data from tropical locations are scarce. Single pregnant females were captured on Cuba in February, April, and May; one of these contained three embryos, whereas the other females had two embryos each (Silva-Taboada, 1979). A lactating bat was captured on Cuba during May, and a postlactating adult was taken in July. On Puerto Rico, a lactating individual carrying three pups was found on 4 June (Rodríguez-Durán, 1999). The mother weighed 7.8 grams, but each offspring already weighed an average of 3.9 grams and was coated with cinnamon-red fur. The large size and furred condition of the young suggest that they were born in mid- to late- May.

In temperate regions, red bats apparently mate in autumn and store sperm during hibernation, thus delaying fertilization until spring. Whether red bats in tropical climates use delayed fertilization is unknown. However, nonhibernating populations of another lasiurine from subtropical Paraguay, the southern yellow bat, copulate about three months before pregnant females can be found, indicating sperm storage (Myers, 1977). Thus, it is possible that red bats on Puerto Rico also dissociate mating and ovulation, in a manner similar to their temperate, hibernating relatives.

Although Shump and Shump (1982) reported a variety of parasites from the red bat, including mites, fleas, batbugs, flukes, and protozoans, no data are available for specimens from Puerto Rico, and the only parasite documented for Cuba is a nycteribiid fly (Silva-Taboada, 1979). Presumably rates of infestation for external parasites are low in the red bat, as they are in the red fig-eating bat, reflecting the solitary lifestyle of these foliage-roosting species (Gannon and Willig, 1995).

FAMILY MOLOSSIDAE
Free-tailed Bats

The family Molossidae is large, containing 16 genera and about 100 species (Simmons, 2005). It has essentially a worldwide distribution, with representatives on every continent except Antarctica. Most species are restricted to areas with tropical climates, but the family is also well represented in temperate areas of southern Europe and Asia as well as North and South America. Two species, each in a different genus, live on Puerto Rico.

Bats in this family are small to medium in size, with body mass ranging from 5 grams in the Mato Grosso dog-faced bat of South America to almost 200 grams in the hairless bat of Malaysia and the Philippines. Fur is most often brown, gray, or black; short in length; and velvetlike in texture. The hairless bats are unique among bats in that adults possess only a coating of fine hairs that are so sparse that the bats appear naked and have the wrinkled look of a newborn (Hutson et al., 2001). Though small, the eyes of molossids are clearly observable. The broad ears point forward, lie against the side of the head, and do not face into oncoming air during flight, thus minimizing aerodynamic drag (Vaughan, 1966). Pinnae commonly are joined by a band of tissue across the forehead. Molossid ears include a tragus, but it is often small and dwarfed by an antitragus, which projects from the base of the pinna lateral to the tragus. Throat (gular) glands typically are present in one sex or both. The scent of their musky secretions may play a role in mating behavior of the bats, but to humans, the distinctive aromas range from "pleasant and fruity" to an "unbearably foul stench," depending on species (Schutt and Simmons, 2002:227). Wing membranes are noticeably thick and leathery compared with those of other bat families in the Antilles. As the common name of the family implies, the tail vertebrae project far beyond the posterior edge of the uropatagium of a free-tailed bat (Fig. 9). Finally, molossids typically have hairy feet, with some bristlelike hairs extending well past the claws.

Molossids have long, narrow wings and are swift and enduring fliers, making them the mammalian counterpart of swifts and swallows (Vaughan, 1966). In free-tailed bats the fifth digit, which extends back from the wrist and supports the middle of the wing membrane, is only about half the length of the forearm, whereas in leaf-nosed bats, which have a broad wing, the fifth digit and forearm are similar in length (Reid, 1997). Fast flight and long, narrow wings are not conducive to foraging in cluttered environments, and not surprisingly, these bats typically search for insects above the forest canopy or in other open spaces. Some, such as the Brazilian free-tailed bat, occasionally forage up to 3,000 meters above the ground (McCracken, 1996). The hind limbs are robust and well muscled compared with those of most other bats, and consequently molossids are surprisingly adept at terrestrial locomotion (Vaughan, 1966). Despite their agility on the ground, these bats have difficulty taking flight when grounded—another consequence of having a wing designed to function best at high speed.

Like many other bats, molossids typically roost in trees, buildings, and caves, but some species rely on more unusual sites. The dwarf dog-faced bat of South America, for example, may occupy decaying logs, and Roberts's flat-headed bat of South Africa is found under slabs of rock (Nowak, 1999). Although many molossids roost as individuals or in small groups of 5–20 animals, some species gather in huge colonies that contain thousands or even millions of individuals. An estimated 20 million Brazilian free-tailed bats—the largest aggregation of mammals in the world—occupy Bracken Cave in Texas (Tuttle, 1994). Molossids do not hibernate, and many temperate populations migrate to warmer climates for winter.

Molossus molossus Pallas, 1766
Velvety Free-tailed Bat
Murciélago de Techos

Taxonomy: Within the New World genus *Molossus,* there are at least five and perhaps eight or more species (Koopman, 1993; López-González and Presley, 2001; Simmons, 2005), and within *Molossus molossus,* there may be up to nine subspecies (Silva-Taboada, 1979). Some of this taxonomic uncertainty revolves around whether mainland subspecies of the velvety free-tailed bat, *M. m. coibensis* and *M. m. aztecus,* are separate species (Dolan, 1989). In any event, we follow Varona (1974) and Pine (1980) and recognize the subspecies *Molossus molossus fortis* as the taxon occurring on Puerto Rico and the Virgin Islands. Type locality for the species is Martinique in the Lesser Antilles, whereas the subspecies description is based on a specimen from Luquillo, in eastern Puerto Rico.

Name: The Greek word *molossus* means "mastiff," and the word refers to

Fig. 37 Portrait of a velvety free-tailed bat.

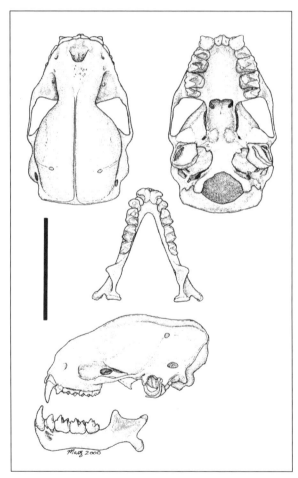

Fig. 38 Skull and mandible of a velvety free-tailed bat
(bar = 10 mm).

the somewhat doglike snout of this species in particular and the family in general (Jaeger, 1955). The subspecific epithet *fortis* comes from the Latin for the word "strong." It is not clear how this term relates to the velvety free-tailed bat, but we speculate that the word refers either to the strong odor produced by these bats or perhaps to the noticeable muscular appearance of the animal when it is held with outstretched wings. Although the name velvety free-tailed bat occurs frequently in the literature, Wilson and Cole (2000) propose use of Pallas's mastiff bat as the common name for this molossid.

Distribution and Status: The velvety free-tailed bat is widespread in the New World, ranging from Oaxaca in Mexico to Uruguay and northern Argentina. It is the only member of the genus to invade the West Indies, and unlike most other species of bat, it likely colonized the islands from South America, migrating northward through the Lesser Antilles (Koopman, 1989). It currently lives on most islands of both the Greater and Lesser Antilles (Map 16). This molossid is found throughout Puerto Rico as well as on Culebra and Vieques; it is most common in areas of human habitation.

Although actual numbers are not available, we believe that the population of velvety free-tailed bats in Puerto Rico is declining. Anthony (1925:67) commented on seeing a "multitude of the little animals . . . pursuing their zigzag courses about the streets." Even in the 1960s, our impression was that these molossids were abundant. Although still widespread today, this species generally goes unnoticed except by the owners of houses where this bat roosts. Reasons for the apparent decline are unknown but may be related to economic growth of the island during the second half of the twentieth century. Such growth was accompanied by an increased use of chemical pesticides, a spectacular expansion of modern urban areas, and a reduction in the number of wooden houses, which are the main roosting sites of large colonies.

Measurements and Dental Formula: Total length is 99–104 mm; length of tail, 36–41 mm; length of hind foot, 8–12 mm; height of ear, 11–13 mm; and length of forearm, 38–40 mm. Body mass is 13–18 g. Dental formula is incisors 1/1, canines 1/1, premolars 1/2, molars 3/3 = 26, which is fewer teeth than are found in any other species of bat on Puerto Rico.

Description: The velvety free-tailed bat is a moderately small bat, weighing

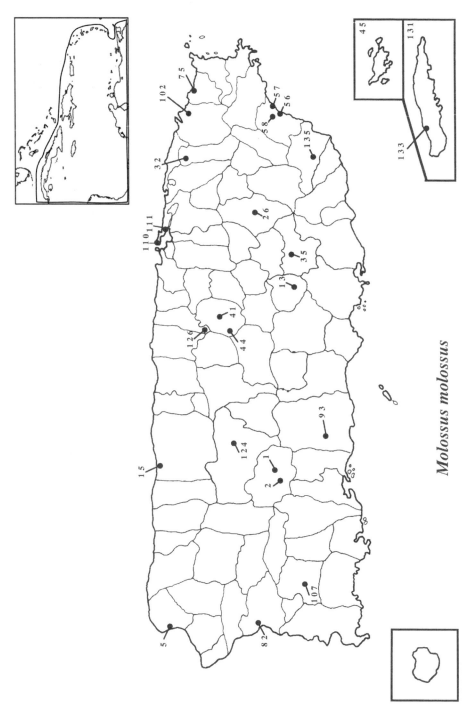

Map 16 Distribution of the velvety free-tailed bat on Puerto Rico and in the Caribbean basin. Location of numbered sites is detailed in Appendix 8.

Molossus molossus

less than 20 grams. The short, rounded ears are connected across the forehead by a thick membrane, and the well-developed antitragus appears almost circular in outline (Fig. 37). The snout is short, broad, and blunt, and the nostrils appear slightly tubular. Males possess a well-developed gular gland. As in other molossids, wings are long and exceedingly narrow, and the tail vertebrae are free, extending beyond the end of the broad tail membrane for approximately half the length of the tail. Conspicuously long hairs project from the feet. Overall, the fur is short, velvety, and dark brown, although the basal portion of each hair is white. The white base is not obvious on the back, unless the hairs are examined closely, but fur is shorter on the belly and chest, which makes the white band more apparent there. Totally white individuals with pinkish skin (albinos) are known from Canóvanas and Utuado (Heatwole et al., 1964).

The most diagnostic aspect of the skull and mandible of a velvety free-tailed bat is number of incisors. Velvety free-tailed bats possess only one pair of upper and one pair of lower incisors (Fig. 38). Although red bats and Brazilian free-tailed bats also have only one pair of upper incisors, a notch in the palate separates the teeth in those species. However, no such gap occurs in the velvety free-tailed bat. The greater bulldog bat also has a single pair of incisors in the mandible, but its skull is much larger than that of a velvety free-tailed bat. Greatest length of the skull is about 27 millimeters in the bulldog bat compared with 17 millimeters in the free-tailed bat.

Natural History: In the early twentieth century, Anthony (1925:67) reported that this species was "almost strictly an inhabitant of old dwellings" on Puerto Rico, and this remains largely true today, even though fairly new buildings may be occupied occasionally. A wooden house with a corrugated metal roof is a common roosting structure for the velvety free-tailed bat; the bat typically enters such a house by crawling under the roofing material, but then it often moves into the walls, where air temperatures are lower. It may also squeeze into cracks in the joints of stone buildings or unfinished areas in the walls of cement houses, and it is commonly found in abandoned military bunkers on Vieques (Rodríguez-Durán, 2002). These bats often occupy manmade bat houses on the Cayman Islands (Band, 2000), and out of 64 roosts found by Silva-Taboada (1979) on Cuba, 52 were in buildings, nine were in cavities in trees (especially royal palms), two were in rock crevices, and one was in a crevice in a wooden electrical pole.

Males are more likely to occupy tree holes, either as solitary individuals or in small bachelor colonies. Both sexes may be found in buildings, although even then, males usually roost apart from the females. Occasionally, as many as 4,000–5,000 velvety free-tailed bats are found in human dwellings on Cuba (Silva-Taboada, 1979), but a colony of 10–20 animals is more typical of this species on Puerto Rico. The bats are usually quiet during the day and seem indifferent to most activity by humans. Although Anthony (1918) indicated that a strong musky odor is commonly associated with buildings inhabited by these bats, we have not noticed this, nor have we heard complaints from homeowners.

The velvety free-tailed bat is crepuscular and has a distinct bimodal pattern of activity (Chase et al., 1991; Silva-Taboada, 1979). On Trinidad, where behavior has been studied in detail, foraging begins between 24 minutes before and 18 minutes after sunset and ends just 30–60 minutes after it begins. After the evening hunt, a period of social activity marked by frequent vocalizations occurs inside the roost. Socializing ends before midnight, with the animals becoming inactive until they leave for a second short feeding bout just before dawn. This bimodal pattern of activity varies only on nights when light rain occurs. Light showers apparently induce a brief flurry of activity by insects, and some bats respond by foraging for short periods of less than 10 minutes, thus taking advantage of the unexpected abundance of food.

During the late-night resting period, bats may drop their body temperature slightly and enter shallow torpor. This energy-saving mechanism also is used during the day when air temperatures fall below 20°C. In Cuba entire colonies remain inactive for days during unusually cold weather (Silva-Taboada, 1979), and we have seen similar behavior in a group of caged bats in Puerto Rico.

The early evening departure of velvety free-tailed bats has potential benefits but also potential drawbacks. These insectivorous bats apparently minimize competition by hunting after diurnal swifts and swallows have roosted for the night, but before the onset of foraging by most other bats (Chase et al., 1991). Nevertheless, the bats do most of their foraging at a time when even humans have enough light to see by, and consequently, these mammals are exposed to predation by late-hunting raptors. The American kestrel is known to attack this molossid in Cuba, and the bat falcon consistently appears during evening emergence in Trinidad (Chase et al., 1991; Silva-Taboada, 1979). Although predation on velvety free-tailed bats on Puerto Rico has not been documented yet, raptors do prey on the sooty mustached bat, another Puerto Rican species that often begins foraging before sunset (Rodríguez-Durán and Lewis, 1985).

Velvety free-tailed bats often follow regular foraging routes, and the short duration of feeding bouts suggests that they do not fly far from the roost. For instance, in western Puerto Rico, we watched a group of these bats commute along the edge of a gorge every evening for over a year. That these bats forage close to home is supported by homing experiments in Cuba, showing that they did not return to the roost when released from distances greater than 20 kilometers (Silva-Taboada, 1979). This maximum homing distance appears shorter than that of most other insect-eating bats in the Antilles.

Diet has been studied through analysis of fecal pellets and stomach contents (Rodríguez-Durán et al., 1993; Silva-Taboada, 1979). On Puerto Rico beetles occurred in 83 percent of 41 pellets that were examined. Moths and true bugs appeared in about half the pellets, whereas flies and ants or wasps occurred in less than 10 percent. On Cuba diet was more diverse and included insects from eight orders. Leafhoppers dominated, appearing in 65 percent of 26 stomachs that were analyzed, with crickets and roaches coming in a distant second at 38 percent. Other types of insect eaten included moths, beetles, true bugs, mayflies, and earwigs. As with many other insect-eating bats, such variation may reflect opportunistic feeding behavior by the velvety free-tailed bat.

This bat hunts in urban or rural areas using echolocation calls that are somewhat unusual (Kössl et al., 1999). Although most echolocating bats emit a train of pulses, each very similar to the next in terms of the frequencies used, the velvety free-tailed bat emits a sequence of paired pulses. Individual pulses are frequency modulated, but the first member of a pair varies in frequency from about 30 to 34 kilohertz and the second from 35 to 40 kilohertz. Use of paired calls of different frequency may allow the bat to process the echo from one call while emitting the next.

Velvety free-tailed bats are seasonally polyestrous, typically giving birth twice each year at specific times (Krutzsch and Crichton, 1985,

1990; Silva-Taboada, 1979). Females first conceive in March and deliver a hairless youngster in June; females mate again soon after young are born (postpartum estrus), and a second birth occurs in September. Not all females in a colony may experience the second pregnancy (Fabian and Vera-Marques, 1989; Silva-Taboada, 1979). Although males have sperm present in the reproductive tract year-round and presumably are capable of mating, females are not receptive until March. The gular gland of a male functions throughout the year, but it greatly hypertrophies on a seasonal basis, becoming most active during the reproductive season of the females. The gland probably plays a role in sexual attraction (contrary to Puerto Rican folklore, which suggests that the gland is somehow used to suck the blood or drain the vitality of humans). As in all molossids, litter size is only one.

Parasites of the velvety free-tailed bat have not been examined yet on Puerto Rico. On Cuba, internal parasites include one tapeworm, two species of fluke, and five kinds of roundworms (Silva-Taboada, 1979). External parasites consist of a multitude of macronyssid, labidocarpid, sarcoptid, myobiid, and rosensteiniid mites, argasid ticks, streblid batflies, and a polyctenid bug.

Tadarida brasiliensis Geoffroy Saint-Hilaire, 1824
Brazilian Free-tailed Bat
Murciélago Viejo

Taxonomy: The Brazilian free-tailed bat is one of 10 species in the genus *Tadarida,* one of the most widespread genera of bats, with representatives on all habitable continents (Simmons, 2005). *T. brasiliensis,* however, is the only species that occurs in the New World. Within *T. brasiliensis* there are nine subspecies, and the smallest of these, *T. b. antillularum,* lives on Puerto Rico. The type locality for *T. brasiliensis* is Paraná, Brazil, whereas *T. b. antillularum* was described originally from a specimen taken at Roseau, on the island of Dominica (Wilkins, 1989).

Name: Although scientific names of many organisms are constructed carefully and have meaning, some biologists consistently invent names that simply sound like Latin. The generic name *Tadarida* seems to be one of these. It was coined by Rafinesque in the early nineteenth century, but

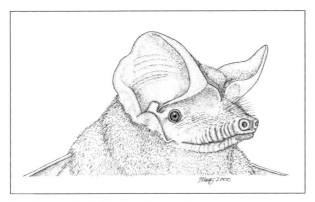

Fig. 39 Portrait of a Brazilian free-tailed bat.

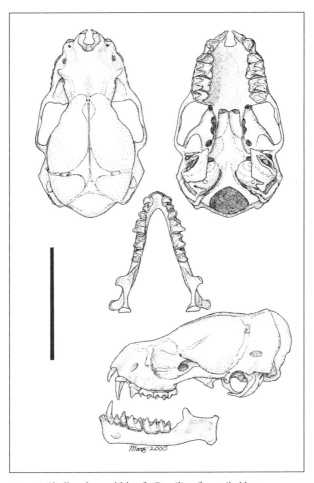

Fig. 40 Skull and mandible of a Brazilian free-tailed bat
(bar = 10 mm).

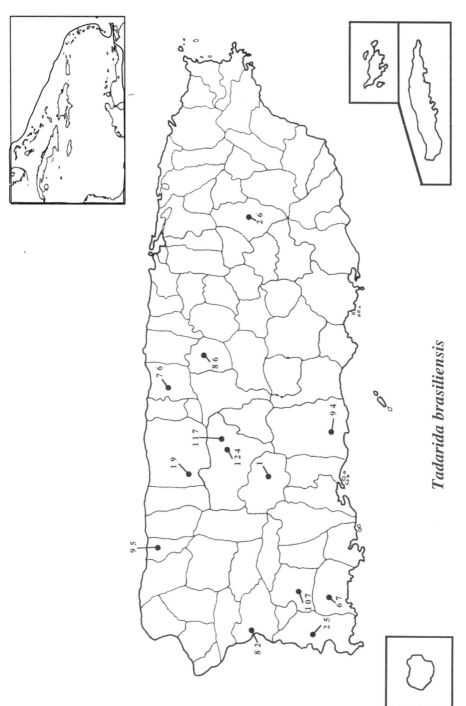

Map 17 Distribution of the Brazilian free-tailed bat on Puerto Rico and in the Caribbean basin. Location of numbered sites is detailed in Appendix 8.

Tadarida brasiliensis

"concerning [the name's] origin and meaning he left us no word" (Jaeger, 1955:255). The specific epithet *brasiliensis* refers to Brazil, where the type was collected, whereas the subspecific designation *antillularum* means "of the Antilles" and indicates where this particular subspecies lives today.

Distribution and Status: The Brazilian free-tailed bat is distributed widely in the New World. In the United States, it lives roughly south of a line stretching from southern Oregon to North Carolina. Its range continues through Mexico and Central America and into northern South America, west of the Andes. Farther south, the range expands eastward to the Atlantic Ocean, and this animal can be found in most of Argentina and Chile as well as southern Brazil. The subspecies *T. b. antillularum* inhabits Puerto Rico and the Lesser Antilles, as far south as St. Lucia (Wilkins, 1989; Map 17).

Populations of the Brazilian free-tailed bat in Mexico and the United States have declined dramatically over the last 50–100 years (McCracken, 1999). A major cause of these declines likely is disturbance or destruction of roost sites containing large colonies, particularly maternity sites, but indirect poisoning by pesticides also may have played a role (Clark, 1981; Clark and Shore, 2001). Largely because of these declines, the IUCN considers the Brazilian free-tailed bat a "near threatened" species (category "LR nt") and has established a Species Action Plan to aid in recovery of the Brazilian free-tailed bat to its former abundance (Hutson et al., 2001). The IUCN plan recommends, among other things, that Mexico and the United States formalize an agreement to protect this species, because many individuals, especially reproductive females, seasonally migrate between the two countries.

In the early twentieth century, Anthony (1925) reported the Brazilian free-tailed bat as one of the most common bats on Puerto Rico. Few colonies are known today, and we believe that it is one of the least abundant species on the island. Reasons for the apparent decline of the island population are unknown, but as with the velvety free-tailed bat, increased use of pesticides and inexorable urban sprawl are possibilities worth investigating.

Measurements and Dental Formula: Total length is 79–98 mm; length of tail, 33–41 mm; length of hind foot, 6–9 mm; height of ear, 8–15 mm;

and length of forearm, 37–41 mm. Body mass is 7–12 g. Dental formula is 1/3, canines 1/1, premolars 2/2, molars 3/3 = 34.

Description: The snout of a Brazilian free-tailed bat is short and rather blunt, and the upper lip has a series of distinct vertical wrinkles (Fig. 39). The rounded ears are wide but short, extending to the tip of the snout or just beyond when gently laid forward. Unlike the ears of a velvety free-tailed bat, those of a Brazilian free-tailed bat do not join on the forehead but remain separated by a small gap of 1–2 millimeters. The antitragus appears half-moon shaped in the Brazilian free-tailed bat but more circular in the velvety free-tailed bat. A gular gland is present in both sexes but reaches maximal development only in mature males (Gutierrez and Aoki, 1973). The brown fur is short and dense and is darker on the dorsum than the venter. The proximal part of the long, narrow wings also is covered partly by hair. As in other molossids, the uropatagium is quite broad, with the tail extending beyond the membrane by up to 20 millimeters, or roughly 50 percent of its length (Silva-Taboada, 1979; Wilkins, 1989). Long, bristly hairs extend well past the tips of the toes. On Puerto Rico, this species is most similar to the velvety free-tailed bat, but the smaller body size and wrinkled upper lip of a Brazilian free-tailed bat make it easy to differentiate these two species.

The skull of a Brazilian free-tailed bat is also easy to recognize. A notch in the front of the palate distinctly separates the upper teeth at the midline (Fig. 40). This notch is narrow at its mouth and widens toward the rear of the indentation. Red and big brown bats are the only other species that have a notch in the palate, but their gap is more U-shaped, with the widest portion near the opening of the notch rather than in the rear.

Natural History: In other parts of its range, the Brazilian free-tailed bat is an extremely common animal and forms the largest concentrations of mammals in the world, with some caves containing millions of individuals (McCracken, 1999). However, colonies on Puerto Rico are comparatively tiny, consisting of only tens or hundreds of bats. The species is somewhat adaptable, and most island colonies roost either in buildings or in small caves (Anthony, 1925; Silva-Taboada, 1979). When using caves the bats typically hang in cool, well-ventilated areas, where they form tight clusters, although the extent of clustering may vary with temperature. Solitary individuals often hide in rock crevices or under roofing tiles (Silva-Taboada, 1979; Wilkins, 1989).

The Mexican subspecies, *T. b. mexicana,* annually undertakes long migrations from Mexico to the southern United States, and the Cuban subspecies, *T. b. muscula,* evidently does not remain year-round in the same roost, suggesting at least a short-distance migration (Silva-Taboada, 1979). On Puerto Rico, however, there is no evidence indicating seasonal movements, although only one colony of Brazilian free-tailed bats has been examined extensively.

This colony occupied an abandoned railroad tunnel near Quebradillas. The tunnel was visited sporadically from 1986 to 1996, except from May 1992 to April 1993, when it was examined twice each month. Between 200 and 300 free-tailed bats were present during each visit, with males and females occurring in approximately equal numbers. Despite the large number of adult females, newborns were never observed, indicating that this site was not used as a maternity roost. Further study of other colonies of Brazilian free-tailed bats on Puerto Rico is needed to determine typical population dynamics at maternity and nonmaternity sites.

The Brazilian free-tailed bat "is easily the best flyer of the island species," having flight that is "under perfect control" and "very rapid" (Anthony, 1925:63). Estimated speed of typical flight is 64 kilometers per hour (Davis et al., 1962). This mammal begins leaving its roost between sunset and 36 minutes after sunset to forage (Silva-Taboada, 1979). Bats from large colonies on the mainland generally leave in tight columns that wind across the countryside, whereas the exodus from smaller colonies, such as those on Puerto Rico, may be more diffuse. These small bats are capable of flying more than 50 kilometers to their foraging grounds and hunting at altitudes up to 3,000 meters, although shorter distances and lesser heights are probably more typical (Davis et al., 1962; McCracken, 1996; Wilkins, 1989). As in other species, an early departure exposes the Brazilian free-tailed bat to predation by visually oriented predators. Although this molossid is preyed upon by a variety of raptors in other parts of its range, including American kestrels and red-tailed hawks (Silva-Taboada, 1979; Wilkins, 1989), we have yet to observe such predation on Puerto Rico.

Brazilian free-tailed bats are aerial insectivores that typically hunt in open spaces, often well above the ground or trees. When searching for prey, these bats emit sounds that are brief and almost constant in frequency (Simmons et al., 1979). Such sounds provide little information other than simple presence or absence of something nearby. However, once an object is detected, the bat transforms its call so that it sweeps from 75 to 40 kilohertz. These frequency-modulated calls during the approach allow the bat to identify the object as well as to determine its location more precisely.

Diet of insular populations of the Brazilian free-tailed bat differs from that of continental populations. For instance, Whitaker et al. (1996) report that diet of *T. b. mexicana* in Texas consists largely of moths (33 percent), beetles (26.4 percent), ants and wasps (20.5 percent), and flies (14.5 percent). This contrasts sharply with data gathered by Whitaker and Rodríguez-Durán (1999) from a colony of *T. b. antillularum* in western Puerto Rico, where flies (36.1 percent), ants and wasps (31.3 percent), and moths (20.1 percent) dominated the diet, with smaller amounts of leafhoppers (5.8 percent), true bugs (3.8 percent), and beetles (2.2 percent) also present. Food habits of *T. b. muscula* from Cuba are similar to those of Brazilian free-tailed bats on Puerto Rico (Silva-Taboada, 1979).

Dietary differences between populations of bats on the continent and those on Caribbean islands may be attributable, at least in part, to differences in communities of bats found in such locations. On Cuba and Puerto Rico, predation on beetles and moths like those consumed by free-tailed bats in Texas is largely the role of the abundant and ubiquitous mormoopids—the ghost-faced and mustached bats. In mainland communities, bats in the vespertilionid genera *Myotis* and *Pipistrellus* typically consume large amounts of flies as well as small moths and beetles. These genera, however, are absent from the Greater Antilles, and it appears that the Brazilian free-tailed bat has moved into this vacant niche, conceding most of the moths and beetles to the mormoopids.

The Brazilian free-tailed bat gives birth to a single offspring once each year, in late spring or early summer, after a gestation of 11–12 weeks (Sherman, 1937; Wilkins, 1989). In Texas, parturition occurs in June or early July, and weaning usually takes place in August (Davis et al., 1962; Kunz and Robson, 1995). In Cuba, pregnant females are found from April through June, with parturition beginning in July and lactation ending in August (Silva-Taboada, 1979). Timing of reproductive events on Puerto Rico is largely unknown, although Anthony (1925) reported near-term fetuses in females captured between 7 and 24 June.

Birth is rapid, with as little as 90 seconds needed to traverse the birth canal, although it takes the newborn an additional 15 minutes to discover a nipple and begin to feed (Sherman, 1937). At parturition, the blind and naked neonates are quite large for a mammal but only average-sized for a bat, weighing about 25 percent of the mother's mass. Females do not roost with their offspring. Instead, a mother deposits her single young within a large cluster of other neonates, and she visits her offspring only a few times each day to nurse (McCracken and Gustin, 1991). Each mother eventually locates her own offspring among the hundreds or thousands of others that are present by relying on spatial memory and the unique sounds and odors produced by her pup. Despite the seeming lack of maternal attentiveness, preweaning mortality is quite low, less than 2 percent. Youngsters begin to fly and forage after about six weeks and reach 90 percent of adult mass by seven weeks (Kunz and Robson, 1995). In late lactation, large pups are nourished by milk containing 28 percent fat—the highest fat content reported for milk of any bat (Kunz et al., 1995). Brazilian free-tailed bats survive up to eight years in the wild (Nowak, 1999).

The Brazilian free-tailed bat is a wide-ranging, often abundant species, making it a favored study animal for biologists, including parasitologists. Consequently, a large number of parasites have been reported throughout its range (Wilkins, 1989), although such a study has never been conducted on Puerto Rico. Internal parasites from Cuba include flatworms, tapeworms, and roundworms; external parasites include chiggers, myobiid and macronyssid mites, and streblid flies (Silva-Taboada, 1979).

Conservation of Bats in the Caribbean Basin

Bats are the only native land mammals on most islands of the Caribbean, and their unique biology poses a number of challenges for successful conservation and management. Even though bats are small mammals, they are long-lived and have low reproductive rates, often producing only one or two offspring each year (Barclay and Harder, 2003; Kurta and Kunz, 1987; Tuttle and Stevenson, 1982). Consequently, bat populations cannot sustain high mortality rates, nor do they recover quickly from sudden population declines (Racey and Entwistle, 2003).

Threats to Bats

The colonial nature of most Caribbean species of bat makes them susceptible to local extinction because the bats are concentrated in few places. A misinformed homeowner, for example, easily can cause the death of 50 or more bats by applying pesticides or using some other inappropriate means to remove bats from a house. Similarly, a group of people entering a cave disturbs not one bat but potentially thousands, and destruction of that cave, or even alteration of its microclimate, may displace hundreds of thousands of animals belonging to multiple species (Arita, 1993; Rodríguez-Durán, 1998).

Degradation of the environment and loss of foraging and roosting habitat are common problems for bats throughout the Neotropics (e.g., Arita and Ortega, 1998; Marinho-Filho and Sazima, 1998), including islands of the Caribbean (e.g., Petit, 1996). Urban expansion, rural development, deforestation, spraying of pesticides, water pollution, and other human activities indirectly have many negative impacts on bats (Clark and Shore, 2001; Hutson et al., 2001; Palmerim and Rodrígues, 1992). On some islands, laws that ostensibly protect wildlife still consider bats as vermin (e.g., St. Vincent Wildlife Protection Act of 1987; Vaughan and Hill, 1996), and this legal misperception probably leads to the needless death of many bats. Even natural events, like Hurricane Hugo or the eruption of the Soufriere Hills Volcano on Montserrat, can have devastating long-term effects on bat populations of Caribbean islands (Pedersen 2000, 2002; Pedersen et al., 1996).

In addition, insular populations generally are more prone to extinction than their mainland counterparts simply because populations on islands are smaller and less likely to be rescued by animals migrating from less affected areas. One species becoming extinct on any one of the islands is termed an extinction event. Biologists are already aware of 69 extinction events involving bats on Caribbean islands (Morgan, 2001; Suárez and Díaz-Franco, 2003; see Table 5), and more surely will occur. On Curaçao, for example, the continued existence of the Trinidadian funnel-eared bat is in doubt, because a recent census indicated that only 57 animals remained (Petit, 1996). On Guadalupe, one of the seven bat species on the island is classified as endangered by the IUCN; three others are considered vulnerable, and the remaining three are described as near threatened (Hutson et al., 2001). Virtually every island group in the Caribbean basin harbors at least one species that is considered near threatened, vulnerable, or endangered by the IUCN.

Why Protect Bats?

Why should citizens and governmental agencies follow these recommendations? There are many reasons, ranging from the practical to the philosophical, for people to accept the duties of being wildlife stewards and to maintain populations of bats for future generations. We have noted the important ecological and economic roles that bats play. Altringham (1996:231), moreover, argues that humans should conserve bats simply because "they have a right to a place on this planet." In addition, Immanuel Kant, one of the greatest philosophers of the eighteenth century, stresses that our duties as human beings toward animals are really indirect duties toward humanity (Heath and Schneewind, 1997:212). "If a man has his dog shot, because it no longer can earn a living for him, he is in by no means in breach of any duty to the dog, since the latter is incapable of judgment, but he thereby damages the kindly and humane qualities in himself, which he ought to exercise in virtue of his duties to mankind. . . . Lest he extinguish such qualities, he must already practice a similar kindliness towards animals."

Many islands of the Caribbean have passed laws to protect the environment in general and caves and wildlife in particular. In Puerto Rico, for example, the Wildlife Bill of 1999 called for establishment of rules to protect native and migratory animals, whereas the Law for the Protection and Conservation of Caves, Caverns, and Sinkholes, passed in 1985, indirectly extended protection to habitats potentially used by bats and other organisms. Even Haiti, one of the most impoverished countries in the region, created a Ministry of the Environment in 1995 to implement a National Environmental Action Plan (Sergile and Woods, 2001).

Mere existence of such laws and bureaucracies, however, does not ensure actions on behalf of the environment, because implementation of environment-friendly policies often is stalled or misdirected by political and economic interests. Ultimately the people of each island and the governmental officials they empower must recognize that conservation "is not an obstacle to progress. It is the guarantee that men will have what is necessary to meet the demands of the future. It is the discovery of strategies that integrate the satisfaction of human needs with development activities that guarantee the permanence of our natural heritage, for this and future generations" (Ortiz, 1989:854).

Recommendations

Rather than witness the decline of a species to a level that requires inclusion on a list of threatened or endangered animals, at which time special measures may be needed to ensure survival, it is better to take steps that maintain adequate numbers before the population dwindles toward extinction. The following simple recommendations are intended to aid in the conservation of bats on Puerto Rico and other islands of the Caribbean.

1. Anyone entering caves or other roosting sites that are inhabited by

bats should avoid disturbing the animals. Frequency of visitation and number of people entering a roost each time should be minimized. Noise within and near the entrance of a bat cave should be minimal as well. Battery-powered lights or chemical lights are preferred, and beams of light should never be directed at roosting bats. No litter that could pollute the soil, water, or air should be left in a cave.

2. Biologists should identify the locations of maternity colonies of bats and determine the reproductive period. During the reproductive period, entrance into a roost should be banned, except for appropriate scientific studies.

3. If disturbance is frequent and uncontrolled, management personnel should explore the use of "gates." Gates typically consist of a network of steel bars that prevent entry by humans but allow bats to fly through unimpeded. Such structures have been used successfully to protect cave-dwelling populations of small-bodied, temperate bats (Tuttle and Taylor, 1998), although the effects of gates on tropical species, especially large-bodied ones or those that exist in huge colonies, are generally unknown. Installation of a gate should be preceded and followed by intensive monitoring to ensure that the well-intentioned gate is not having a detrimental effect on one or more species (e.g., restricting access and/or increasing predation). Gating, or any other alteration to the cave entrance, should be considered only after consultation with specialists.

4. Research on bats should be conducted only by trained individuals who are familiar with proper techniques for working with bats. Harp traps and mist nets are suitable for capturing bats without harm (Kunz and Kurta, 1988), but such devices should be stationed outside a roost to minimize disturbance. In addition, netting near caves or other roosts containing a large number of bats should be done cautiously to prevent many bats from becoming entangled at one time and overwhelming the researchers' ability to extract the animals without harm. Use of hand nets inside a roost should be limited. After capture, one should handle and transport the bats so as to ensure a minimum of discomfort, with special attention to providing water and a proper thermal environment.

5. Placement of artificial structures (bat houses) for use by roosting bats should be encouraged as a way to promote coexistence of building-dwelling bats and homeowners who wish to evict them (Tuttle and Hensley, 1993).

Research is needed to determine the size, shape, degree of solar exposure, and type of construction materials that are appropriate for the Caribbean environment.

6. Many species of bat will roost in expansion joints and other crevices in bridges, especially when natural crevices are not available (Keeley and Tuttle, 1999). Local highway departments should consider adoption of bat-friendly designs for new bridges and possible retrofitting of existing bridges, as a means of maintaining or creating roosting habitat. Bridges with low traffic volumes may be preferable so that roosting bats are exposed to a lower level of contaminants from passing automobiles and trucks.

7. Governmental and private organizations should consider the needs of bats when reforesting private and public lands, such as parks and roadsides, and in managing existing forests. A diversity of shrubs, trees, and vines that are used by bats for food (Appendix 1) or shelter should be encouraged.

8. The number and extent of forest reserves (Map 3b) should be expanded to provide an abundance of foraging and roosting habitat.

Resources

One of the main incentives for conservation of any species is its popularity, and "threats to bats are often related to ignorance of their lifestyles and roles in ecosystem maintenance" (Hutson et al., 2001:56). Consequently, government agencies, schools, private organizations, and wildlife specialists should encourage public education about the uniqueness of bats, their ecological importance, and the benefits that they provide to humans.

Bat Conservation International is an extremely helpful organization in this regard and distributes a wealth of educational materials concerning bats, including pamphlets, slide shows, and videos, many of which are in both English and Spanish. Although the group Eurobats has a European orientation, this organization provides links to over 30 bat-oriented web sites worldwide. Web sites specific to bats of the Caribbean do not yet exist, but information concerning biodiversity and conservation in the region—topics of direct and indirect importance to bats—may be gleaned from sites maintained by the Caribbean Conservation Association and the Island Resource Foundation. Information specific to activities on Puerto Rico is given on sites operated by the Puerto Rico Conservation Trust, Puerto Rico

Conservation Foundation, Citizens of the Karst, and Inter American University, Bayamón Campus. The Internet addresses for these organizations are:

Bat Conservation International
    http://www.batcon.org
Eurobats
    http://www.eurobats.org
Caribbean Conservation Association
    http://www.ccanet.net/index2.shtml
Island Resource Foundation
    http://www.irf.org
Puerto Rico Conservation Trust
    http://www.fideicomiso.org
Puerto Rico Conservation Foundation
    http://www.tld.net/users/fconserv
Citizens of the Karst
    http://www.cdk-pr.org
Inter American University, Bayamón Campus
    http://bc.inter.edu

The thirteen species of that occur on Puerto Rico belong to five different families. All species can be identified easily using a dichotomous key offering choices between two alternative sets of characteristics. By selecting the most appropriate set, the reader is led in a step-by-step process to final identification of the specimen.

The first key is based on external features of living bats, and the second key uses dental and cranial characteristics. Be sure to consult the glossary for terms that may be unfamiliar. Both keys are intended for use with adult specimens. Only a metric ruler is needed to help identify living bats, but a hand lens or dissecting microscope and a ruler or calipers are needed to work with the smaller skulls. Once identification tentatively is made, compare the specimen with the complete description of the species given in the text and with the drawings of it.

Key to Adult Bats Using External Characters

1a. Tail membrane poorly developed, narrow, appearing as an inverted U or V, when legs are spread . . . .
. . . . **Family Phyllostomidae, leaf-nosed bats** 2

1b. Tail membrane well developed, extending entirely across space between legs 6

2a. Snout a truncated cone, appearing piglike; no obvious nose-leaf; tail vertebrae short, hidden in base of tail membrane . . . . Antillean fruit bat, *Brachyphylla cavernarum*

2b. Snout not piglike; nose-leaf obviously present but may be small; tail variable 3

3a. White spots on shoulders . . . . . . Red fig-eating bat, *Stenoderma rufum*

3b. No white spots on shoulders 4

4a. Length of forearm greater than 52 mm . . . . Jamaican fruit bat, *Artibeus jamaicensis*

4b. Length of forearm less than 52 mm 5

5a. Length of forearm greater than 42 mm . . . . .Brown flower bat, *Erophylla sezekorni*

5b. Length of forearm less than 42 mm . . . . .Greater Antillean long-tongued bat, *Monophyllus redmani*

6a. Length of forearm much greater than 55 mm; hind legs and feet disproportionately large . . . .**Family Noctilionidae, bulldog bats:** greater bulldog bat, *Noctilio leporinus*

6b. Length of forearm less than 55 mm; legs and feet normal 7

7a. At least 40 percent of tail extends beyond posterior edge of tail membrane . . . . **Family Molossidae, free-tailed bats** 8

7b. Tail as long as or shorter than tail membrane 9

8a. Upper lip wrinkled; 1–2 mm gap between ears on forehead . . . . Brazilian free-tailed bat, *Tadarida brasiliensis*

8b. Upper lip not wrinkled; ears continuous across forehead . . . . Velvety free-tailed bat, *Molossus molossus*

9a. Tail as long as tail membrane, extending to posterior margin of membrane or slightly beyond . . . .

. . . . **Family Vespertilionidae, plain-nosed bats** 10

9b. Tail shorter than tail membrane, posterior half of tail protrudes from dorsal surface of tail membrane at rest, but totally enclosed in tail membrane when legs spread . . . .

. . . **Family Mormoopidae, ghost-faced and mustached bats** 11

10a. Fur reddish; at least anterior third of tail membrane furred . . . .

. . . . Red bat, *Lasiurus borealis*

10b. Fur brown; tail membrane naked . . . .

. . . . Big brown bat, *Eptesicus fuscus*

11a. Ears broad, rounded, joined across top of head; eye appears to be inside funnel leading to external ear opening . . . .

. . . . Antillean ghost-faced bat, *Mormoops blainvillii*

11b. Ears pointed, not joined across top of head; eyes normal 12

12a. Length of forearm greater than 44 mm . . . .

. . . . Parnell's mustached bat, *Pteronotus parnellii*

12b. Length of forearm less than 44 mm . . . .

. . . . Sooty mustached bat, *Pteronotus quadridens*

Key to Adult Bats Using Cranial and Dental Characters

1a. Incisive foramina present 2

1b. Incisive foramina absent 7

2a. Skull appears L-shaped in lateral view, with forehead and rostrum at an angle of about 90 degrees to each other . . . .

. . . . . .Family Mormoopidae, ghost-faced and mustached bats, in part:
Antillean ghost-faced bat, *Mormoops blainvillii*

2b. Skull does not appear L-shaped in lateral view, with forehead and rostrum at an angle much less than 90 degrees to each other . . . .

. . . . Family Phyllostomidae, leaf-nosed bats 3

3a. Four upper cheek teeth; greatest length of skull more than 26 mm ...
..........................Jamaican fruit bat, *Artibeus jamaicensis*
3b. Five upper cheek teeth; skull length variable 4

4a. Greatest length of skull more than 26 mm ....
...............Greater Antillean fruit bat, *Brachyphylla cavernarum*
4b. Greatest length of skull less than 26 mm 5

5a. Rostrum much wider than it is long ....
.........................Red fig-eating bat, *Stenoderma rufum*
5b. Rostrum obviously longer than it is wide 6

6a. Pinhole-sized foramen present in palate between incisive foramina and incisors (may be covered in incompletely cleaned skulls); upper incisors about equal in size; first two upper cheek teeth (premolars) with distinct, high, narrow middle cusp that is lacking on posterior upper cheek teeth (molars); distinct gap between right and left incisors on mandible ....
..........Greater Antillean long-tongued bat, *Monophyllus redmani*
6b. Pinhole-sized foramen not present in palate between incisive foramina and incisors (may be covered in incompletely cleaned skulls); outer incisors on upper jaw much smaller than inner; cusps on first two upper cheek teeth not noticeably different in height in lateral view from posterior cheek teeth; no distinct gap between right and left incisors on mandible ....
.... Brown flower bat, *Erophylla sezekorni*

7a. Left and right incisors separated by distinct gap in anterior part of palate 8
7b. No distinct gap in anterior edge of palate 10

8a. Gap in anterior edge of palate is U-shaped, with widest portion near front of gap . . . . Family Vespertilionidae, plain-nosed bats 9
8b. Gap in anterior edge of palate is irregular in shape, with widest portion near posterior part of gap . . . . .Family Molossidae, free-tailed bats, in part: Brazilian free-tailed bat, *Tadarida brasiliensis*

9a. One upper incisor; five upper cheek teeth, the first minute and hidden in lateral view . . . . .Red bat, *Lasiurus borealis*
9b. Two upper incisors; four upper cheek teeth . . . . Big brown bat, *Eptesicus fuscus*

10a. One upper incisor . . . . .Family Molossidae, free-tailed bats, in part: velvety free-tailed bat, *Molossus molossus*
10b. Two upper incisors (although one may be minute) 11

11a. Four upper cheek teeth; greatest length of skull more than 24 mm . . . . .Family Noctilionidae, bulldog bats: greater bulldog bat, *Noctilio leporinus*
11b. Five upper cheek teeth; greatest length of skull less than 24 mm . . . . Family Mormoopidae, ghost-faced and mustached bats, in part . . . 12

12a. Greatest length of skull 18 mm or more . . . . Parnell's mustached bat, *Pteronotus parnellii*
12b. Greatest length of skull less than 18 mm . . . . Sooty mustached bat, *Pteronotus quadridens*

Appendices

Appendix 1

Plants Used as Food by Bats

Scientific name[1]	English name	Spanish name	Type of food	Source[2]
Acacia farnesiana	Casha	Aroma	Fruit	
Albizia lebbek*	Woman's tongue	Acacia amarillo	Nectar/pollen	7
Albizia procera*	White siris	Albicia	Nectar/pollen	7
Andira inermis*	Angelin	Moca	Fruit	1, 5
Annona squamosa	Sweet sop	Anón	Fruit	
Calophyllum calaba*	Maria	Maria	Fruit/leaves	1, 2, 5
Carica papaya*	Papaya	Papaya	Fruit	1, 5
Cecropia schreberiana*	Trumpet tree	Yagrumo hembra	Fruit	1, 2, 5, 6, 8
Ceiba petandra*	Silk cotton	Ceiba	Nectar/pollen	9
Chrysalidocarpus lutescens*	Golden-fruited palm	Palma areca	Fruit	1
Chrysobalanus icaco	Coco-plum	Icaco	Fruit	
Chrysophyllum caimito	Star-apple	Caimito	Fruit	
Cinnamomum elongatum*	Laurel bobo	Laurel bobo	Fruit	9
Citrus aurantium	Sour orange	Naraja	Fruit	
Clusea rosea*	Pitch apple	Cupey	Fruit	1
Coccoloba uvifera*	Sea grape	Uva de playa	Fruit	
Colubrina arborescens*	Greenheart	Abeyuelo	Fruit	5
Erythrina poeppigiana*	Mountain immortelle	Bucaré	Leaves	1, 2, 5

Appendix 1 (Continued)

Scientific name[1]	English name	Spanish name	Type of food	Source[2]
*Ficus benjamina**	Benjamin fig	Laurel benjamín	Fruit	9
*Ficus citrifolia**	Jag	Laurel	Fruit	1, 5
Ficus religiosa	Botree	Alamo	Fruit	
Ficus spp.*	Fig	Ficus	Leaves	2
*Ficus trigonata**	Fig	Ficus	Fruit	5
Guazuma ulmifolia	Bastard cedar	Guácima	Fruit	
Hibiscus elatus	Blue mahoe	Majagua	Fruit	
Kigelia pinnata	Sausage tree	Palo de salchichón	Nectar/pollen	
Leptocereus sp.	Columnar cactus	Sebucan	Nectar/pollen	
*Leucaena leucocephala**	Wild tamarind	Zarcilla	Nectar/pollen	7
*Maga grandiflora**	Maga wood	Maga	Fruit, nectar/pollen	5, 7
*Mangifera indica**	Mango	Mangó	Fruit	1
*Manilkara bidentata**	Bullet-wood	Ausubo	Fruit	8
*Manilkara zapota**	Sapodilla	Níspero	Fruit	9
Mastichodendron foetidissimum	Mastic-bully	Tortugo amarillo	Fruit	
*Melicoccus bijugatus**	Spanish lime	Quenepa	Fruit	5
Musa sp.*	Banana	Guineo	Nectar/pollen	4, 7
*Mutingia calabura**	Panama berry	Capulín	Fruit	5, 7
*Passiflora rubra**	Passion flower	Flor de passión	Nectar/pollen	9
Persea americana	Avocado	Aguacate	Fruit	

*Pilosocereus royenii**	Columnar cactus	Sebucán	Nectar/pollen	3
*Piper aduncum**	Elder	Higuillo	Fruit	1, 2, 5, 7, 8
*Piper glabrescens**	Elder	Higuillo	Fruit	8
*Piper hispidum**	Elder	Higuillo	Fruit	8
*Piper marginatum**	Higuillo oloroso	Higuillo oloroso	Fruit	9
Pouteria sapota	Red mammee	Zapote	Fruit	
*Prestoea montana**	Sierra palm	Palma de sierra	Nectar/pollen	8
*Psidium guajava**	Guava	Guayaba	Fruit, Nectar/pollen	7, 9
Rhizophora mangle	Red mangrove	Mangle rojo	Fruit	
*Roystonea sp.**	Royal palm	Palma real	Nectar/pollen	9
Selenicereus sp.	Queen of the night	Reina de la noche	Nectar/pollen	
*Solanum torvum**	Turkey berry	Berenjena cimarrona	Fruit	2, 5, 7
*Spondias dulsis**	Golden apple	Jobo de la India	Fruit	1
*Spondias mombin**	Hogplum	Jobo	Fruit	1, 2
Spondias purpurea	Jamaica plum	Jobillo	Fruit	
*Stenocereus griseus**	Columnar cactus	Sebucán	Nectar/pollen	3
Syzygium jambos	Rose apple	Pomar rosa	Fruit	
*Terminalia catappa**	Almond	Almendro	Fruit	1, 5

[1]An asterisk (*) indicates a plant on which bats on Puerto Rico actually are known to feed. Plant names without an asterisk and lacking an indicated source are documented foods for bats on Cuba (Silva-Taboada, 1979); these plants grow on Puerto Rico, and further study likely will show that they are used by Puerto Rican bats as well.

[2]Sources are: 1 = Diáz-Diáz (1983); 2 = Kunz and Diaz (1995); 3 = Rivera-Marchand (2001); 4 = Rodríguez-Durán and Kunz (2001); 5 = Rodríguez-Durán and Vázquez (2001); 6 = Scogin (1982); 7 = Soto-Centeno (2004); 8 = Willig and Gannon (1996); and 9 = unpublished observations of the authors.

Appendix 2
Body Measurements[1]

Species	Sex	n	Total length	Length of tail	Length of hind foot	Height of ear	Body mass	Length of forearm
Artibeus jamaicensis	M	10	84.4 ± 1.8 (81.0–88.0)	—	15.8 ± 0.9 (15.0–17.0)	21.5 ± 0.8 (20.0–23.0)	45.8 ± 2.2 (43.0–49.0)	60.6 ± 1.6 (58.0–63.0)
	F	10	85.8 ± 2.1 (83.0–89.0)	—	15.4 ± 1.0 (15.0–18.0)	21.9 ± 1.0 (21.0–24.0)	45.3 ± 4.1 (40.0–53.0)	61.5 ± 1.4 (59.0–64.0)
Brachyphylla cavernarum	M	10	90.9 ± 3.3 (86.0–95.0)	—	18.3 ± 1.1 (17.0–20.0)	22.3 ± 1.1 (21.0–24.0)	45.1 ± 4.7 (36.0–52.5)	65.3 ± 1.8 (61.0–67.0)
	F	10	88.5 ± 2.8 (85.0–94.0)	—	19.5 ± 1.0 (18.0–21.0)	22.2 ± 1.1 (20.0–24.0)	43.3 ± 2.2 (40.0–47.5)	65.0 ± 1.9 (63.0–68.0)
Eptesicus fuscus	M	10	115.6 ± 4.8 (109.0–124.0)	45.8 ± 2.5 (43.0–50.0)	11.1 ± 2.0 (43.0–50.0)	17.2 ± 0.9 (15.0–18.0)	15.3 ± 1.6 (13.5–19.0)	48.9 ± 1.4 (47.0–51.0)
	F	10	116.7 ± 4.9 (110.0–126.0)	46.6 ± 7.0 (35.0–70.0)	11.8 ± 1.6 (8.0–14.0)	16.4 ± 2.0 (14.0–20.0)	16.3 ± 3.3 (11.0–20.5)	48.0 ± 2.3 (45.0–51.0)
Erophylla sezekorni	M	10	83.2 ± 1.4 (80.0–85.0)	14.3 ± 0.9 (13.0–15.0)	14.8 ± 0.4 (14.0–15.0)	18.7 ± 0.5 (18.0–19.0)	18.3 ± 1.3 (16.5–20.5)	49.1 ± 1.1 (47.0–51.0)
	F	10	85.6 ± 2.5 (82.0–90.0)	14.7 ± 1.2 (13.0–17.0)	14.2 ± 0.6 (13.0–15.0)	17.8 ± 0.8 (20.0–23.0)	18.3 ± 1.4 (16.0–20.5)	48.0 ± 0.9 (47.0–49.0)
Lasiurus borealis	F	3	100.3 ± 3.2 (98.0–104.0)	43.7 ± 3.5 (40.0–47.0)	8.4 ± 1.9 (6.5–10.0)	12.9 ± 2.4 (9.5–15.0)	8.3 ± 2.2 (6.5–11.5)	42[2]
Molossus molossus	M	5	97.3 ± 3.8 (91.5–102.0)	34.2 ± 1.6 (36.0–39.0)	8.5 ± 2.1 (6.5–12.0)	10.1 ± 1.7 (9.0–13.0)	14.1 ± 1.2 (13.0–16.0)	—
	F	10	101.2 ± 1.4 (99.0–104.0)	38.9 ± 1.7 (36.0–41.0)	8.0 ± 0.0 (8.0)	11.6 ± 0.7 (11.0–13.0)	14.7 ± 1.4 (13.0–17.5)	39.3 ± 0.6 (38.5–40.0)[3]
Monophyllus redmani	M	10	66.2 ± 2.3 (62.0–69.0)	9.4 ± 0.7 (8.0–10.0)	11.0 ± 0.7 (10.0–12.0)	13.4 ± 0.7 (12.0–14.0)	9.1 ± 1.1 (6.0–9.5)	36.9 ± 0.7 (36.0–38.0)
	F	10	64.6 ± 2.8 (58.0–67.0)	8.4 ± 0.5 (8.0–9.0)	11.1 ± 0.3 (11.0–12.0)	12.9 ± 1.0 (11.0–14.0)	8.7 ± 0.9 (6.5–9.5)	36.4 ± 0.7 (36.0–38.0)

Species	Sex	n						
Mormoops blainvillii	M	10	85.1 ± 2.4 (80.0–88.0)	28.4 ± 2.3 (23.0–31.0)	7.3 ± 1.2 (6.0–9.0)	14.5 ± 1.7 (12.0–18.0)	9.7 ± 0.3 (9.0–10.0)	47.8 ± 1.5 (46.0–50.0)
	F	10	83.6 ± 2.8 (78.5–87.0)	27.2 ± 2.8 (21.5–30.0)	8.6 ± 1.3 (7.0–10.0)	13.6 ± 1.3 (11.0–15.0)	9.2 ± 0.6 (8.5–10.5)	48.0 ± 0.8 (47.0–49.0)
Noctilio leporinus	M	10	125.2 ± 3.8 (119.0–130.0)	28.7 ± 2.0 (25.0–32.5)	32.1 ± 1.7 (30.0–34.0)	26.2 ± 2.1 (22.0–29.0)	68.1 ± 5.0 (61.0–77.0)	87.7 ± 1.4 (85.7–90.0)
	F	10	116.9 ± 3.4 (112.0–122.0)	25.2 ± 2.1 (23.0–30.0)	32.1 ± 2.2 (26.0–33.0)	28.2 ± 1.3 (25.0–30.0)	53.5 ± 3.8 (48.0–60.0)	86.9 ± 2.3 (85.4–88.7)
Pteronotus parnellii	M	11	81.5 ± 2.3 (79.0–86.0)	19.6 ± 5.1 (13.0–22.0)	10.2 ± 1.4 (9.0–13.0)	19.6 ± 3.0 (16.0–25.0)	12.4 ± 1.0 (10.0–13.5)	51.9 ± 1.0 (50.0–53.0)
	F	9	84.2 ± 3.8 (81.0–91.0)	18.4 ± 1.7 (15.0–20.0)	10.9 ± 1.9 (8.0–13.0)	21.3 ± 2.3 (16.0–22.0)	14.3 ± 2.1 (11.0–17.5)	54.0 ± 2.6 (52.0–57.0)[3]
Pteronotus quadridens	M	8	63.5 ± 3.5 (59.0–69.0)	19.9 ± 1.1 (18.0–21.0)	8.1 ± 1.1 (6.0–9.0)	15.6 ± 1.0 (14.5–17.0)	6.2 ± 3.5 (4.0–14.5)	38.5 ± 1.0 (37.0–40.0)[4]
	F	10	68.2 ± 5.1 (61.5–80.0)	17.4 ± 2.7 (13.0–21.0)	9.2 ± 1.5 (7.0–11.0)	16.0 ± 1.7 (13.0–19.0)	5.7 ± 0.9 (4.5–7.0)	38.9 ± 0.7 (38.0–40.0)
Stenoderma rufum	M	8	64.4 ± 3.9 (60.0–69.0)	—	14.8 ± 2.1 (12.0–17.0)	19.3 ± 1.8 (15.0–20.0)	23.4 ± 4.0 (20.0–31.0)	48.9 ± 1.0 (48.0–51.0)[4]
	F	4	66.0 ± 3.8 (61.0–69.0)	—	16.8 ± 1.3 (15.0–18.0)	18.8 ± 2.5 (15.0–20.0)	23.7 ± 4.0 (20.0–28.0)	50.2 ± 1.1 (48.0–52.0)[4]
Tadarida brasiliensis	M	10	90.5 ± 3.2 (86.0–98.0)	36.1 ± 2.4 (33.0–40.0)	6.7 ± 0.5 (6.0–7.0)	14.4 ± 0.5 (14.0–15.0)	10.4 ± 0.9 (9.5–11.5)	39.0 ± 0.9 (37.0–40.0)
	F	10	85.4 ± 3.1 (79.0–90.0)	33.9 ± 2.7 (31.0–39.5)	7.3 ± 0.8 (6.0–8.5)	10.8 ± 1.0 (8.5–12.0)	7.8 ± 1.0 (7.0–9.0)	39.1 ± 1.2 (37.5–41.0)

[1] Standard measures of body size in male and female bats from Puerto Rico. Lengths of forearm were measured on live specimens, but all other measurements were taken from tags accompanying specimens from Puerto Rico in the Texas Tech Museum. Linear measurements are in millimeters, and body mass is in grams. Values are given as mean ± standard deviation (range).

[2] n = 1
[3] n = 3
[4] n = 10

Appendix 3
Cranial Measurements[1]

Species	Sex	n	Greatest length of skull	Condylobasal length bone	Length of maxillary toothrow	Length of maxillary
Artibeus jamaicensis	M	15	28.37 ± 0.44 (27.57–29.16)	25.18 ± 0.30 (24.61–25.57)	10.00 ± 0.10 (9.83–10.24)	8.28 ± 0.12 (7.97–8.43)
	F	15	28.45 ± 0.61 (27.49–29.77)	25.23 ± 0.35 (24.53–25.74)	9.90 ± 0.18 (9.62–10.28)	8.17 ± 0.18 (7.97–8.67)
Brachyphylla cavernarum	M	15	31.70 ± 0.46 (30.88–32.49)	28.26 ± 0.52 (27.29–29.25)	10.88 ± 0.19 (10.51–11.15)	8.85 ± 0.21 (8.42–9.37)
	F	15	31.13 ± 0.31 (30.58–31.77)	27.82 ± 0.37 (27.32–28.47)	10.78 ± 0.15 (10.46–11.15)	8.79 ± 0.15 (8.54–9.01)
Eptesicus fuscus	M	8	19.48 ± 0.33 (19.08–20.14)	17.72 ± 0.35 (17.24–18.21)	7.15 ± 0.12 (6.95–7.31)	5.72 ± 0.08 (5.61–5.87)
	F	11	19.79 ± 0.33 (19.28–20.25)	18.14 ± 0.39 (17.56–18.55)	7.28 ± 0.19 (6.85–7.52)	5.28 ± 0.12 (5.61–5.94)
Erophylla sezekorni	M	15	24.69 ± 0.42 (23.91–25.43)	22.36 ± 0.29 (21.79–23.03)	7.93 ± 0.51 (6.19–8.42)	6.32 ± 0.20 (6.09–6.77)
	F	15	24.44 ± 0.36 (23.83–25.03)	22.18 ± 0.28 (21.58–22.58)	8.00 ± 0.17 (7.81–8.49)	6.39 ± 0.20 (6.07–6.93)
Lasiurus borealis	M	0				
	F	2	13.01[2]	11.96[2]	4.50 ± 0.05 (4.46–4.53)	3.61 ± 0 (3.61)
Molossus molossus	M	1	17.13	15.2	6.3	5.07
	F	9	17.11 ± 0.30 (16.64–17.61)	15.09 ± 0.20 (14.80–15.34)	6.18 ± 0.09 (6.01–6.26)	4.95 ± 0.07 (4.83–5.02)
Monophyllus redmani	M	15	19.98 ± 0.28 (19.51–20.45)	18.41 ± 0.24 (18.11–18.89)	6.90 ± 0.12 (6.65–7.10)	5.49 ± 0.20 (5.05–5.72)
	F	15	19.95 ± 0.27 (19.37–20.40)	18.47 ± 0.24 (17.89–18.90)	6.96 ± 0.11 (6.77–7.14)	5.41 ± 0.16 (5.19–5.66)
Mormoops blainvillii	M	12	14.41 ± 0.26 (13.91–14.84)	13.37 ± 0.18 (13.03–13.62)	7.69 ± 0.09 (7.57–7.80)	6.20 ± 0.12 (6.01–6.36)
	F	1	14.05	13.01	6.17	6.17
Noctilio leporinus	M	3	27.70 ± 1.28 (26.68–29.14)	24.69 ± 0.78 (23.85–25.38)	10.53 ± 0.20 (10.30–10.65)	8.83 ± 0.18 (8.65–9.00)
	F	3	26.62 ± 0.46 (26.21–27.11)	23.60 ± 0.22 (23.36–23.78)	10.13 ± 0.12 (10.04–10.27)	8.63 ± 0.08 (8.55–8.70)

Zygomatic width	Width of postorbital constriction	Width of braincase	Width across upper molars	Width across upper canines	Greatest length of mandible
17.06 ± 0.32	7.23 ± 0.22	12.88 ± 0.25	12.54 ± 0.21	8.05 ± 0.16	18.47 ± 0.26
(16.38–17.78)	(6.82–7.51)	(12.41–13.20)	(12.15–12.87)	(7.80–8.28)	(18.14–18.90)
16.98 ± 0.36	7.23 ± 0.15	12.72 ± 0.29	12.34 ± 0.17	7.82 ± 0.16	18.49 ± 0.37
(16.31–17.50)	(6.94–7.46)	(12.11–13.09)	(12.05–12.62)	(7.47–8.08)	(17.70–19.33)
16.97 ± 0.29	6.63 ± 0.19	12.99 ± 0.24	11.08 ± 0.23	7.23 ± 0.17	20.78 ± 0.33
(16.26–17.33)	(6.34–6.90)	(12.57–13.53)	(10.57–11.32)	(6.97–7.53)	(20.28–21.48)
16.76 ± 0.39	6.53 ± 0.15	12.79 ± 0.19	11.00 ± 0.22	7.01 ± 0.14	20.47 ± 0.28
(16.21–17.45)	(6.15–6.78)	(10.48–11.36)	(6.74–7.25)	(6.74–7.26)	(19.90–20.92)
12.79 ± 0.21	4.38 ± 0.09	8.67 ± 0.14	8.01 ± 0.19	5.89 ± 0.17	14.09 ± 0.38
(12.47–12.97)	(4.23–4.48)	(8.52–8.90)	(7.76–8.34)	(5.64–6.13)	(13.38–14.61)
12.90 ± 0.45	4.42 ± 0.15	8.88 ± 0.29	8.19 ± 0.19	6.06 ± 0.12	14.41 ± 0.32
(12.45–13.81)	(4.20–4.66)	(8.48–9.45)	(7.93–8.45)	(5.86–6.24)	(13.78–14.95)
11.59 ± 0.28	4.75 ± 0.14	10.29 ± 0.22	6.83 ± 0.25	5.34 ± 0.18	15.85 ± 0.19
(10.91–11.96)	(4.95–5.03)	(9.98–10.68)	(6.38–7.17)	(5.02–5.70)	(15.44–16.17)
11.44 ± 0.18	4.75 ± 0.11	10.07 ± 0.17	6.82 ± 0.16	5.18 ± 0.19	15.82 ± 0.19
(11.10–11.68)	(4.44–4.88)	(9.79–10.32)	(6.59–7.09)	(4.93–5.57)	(15.49–16.15)
9.35[2]	4.40 ± 0.16	7.18[2]	6.37 ± 0.35	5.13 ± 0.28	8.90 ± 0.08
	(4.28–4.51)		(6.12–6.62)	(4.93–5.32)	(8.84–8.96)
10.76	3.74	8.78	7.76	4.44	11.25
10.55 ± 0.18	3.62 ± 0.07	8.67 ± 0.08	7.83 ± 0.10	4.37 ± 0.06	11.18 ± 0.24
(10.41–10.79)	(3.45–3.68)	(8.50–8.75)	(7.73–8.01)	(4.26–4.42)	(10.88–11.50)
8.50 ± 0.15	3.93 ± 0.10	8.44 ± 0.15	4.77 ± 0.14	3.38 ± 0.09	12.57 ± 0.18
(8.13–8.67)	(3.75–4.09)	(8.16–8.62)	(4.56–5.00)	(3.18–3.58)	(12.09–12.87)
8.26 ± 0.10	3.99 ± 0.08	8.38 ± 0.15	4.74 ± 0.12	3.20 ± 0.07	12.52 ± 0.24
(8.08–8.42)	(3.86–4.11)	(8.17–8.59)	(4.54–5.02)	(3.09–3.37)	(12.36–12.86)
8.61 ± 0.20	4.52 ± 0.07	7.46 ± 0.16	6.19 ± 0.11	4.46 ± 0.09	12.02 ± 0.21
(8.20–8.98)	(4.41–4.64)	(7.16–7.76)	(6.01–6.33)	(4.33–4.58)	(11.52–12.32)
8.50	4.47	7.49	6.06	4.40	11.96
19.94 ± 0.75	7.35 ± 0.19	14.07 ± 0.36	13.08 ± 0.29	9.34 ± 0.36	18.68 ± 0.55
(19.07–20.45)	(7.13–7.49)	(13.71–14.42)	(12.80–13.37)	(8.94–9.65)	(18.06–19.10)
19.07 ± 0.13	7.22 ± 0.09	13.80 ± 0.24	12.61 ± 0.10	8.79 ± 0.08	17.96 ± 0.03
(18.94–19.19)	(7.12–7.30)	(13.52–13.97)	(12.54–12.72)	(8.72–8.87)	(17.93–17.99)

(continued)

Species	Sex	n	Greatest length of skull	Condylobasal length	Length of maxillary bone	Length of maxillary toothrow
Pteronotus parnellii	M	17	20.47 ± 0.19 (20.14–20.75)	19.06 ± 0.17 (18.75–19.36)	8.73 ± 0.14 (8.47–8.96)	7.02 ± 0.10 (6.85–7.23)
	F	13	20.26 ± 0.16 (20.08–20.65)	18.85 ± 0.15 (18.67–19.18)	8.63 ± 0.13 (8.45–8.84)	7.01 ± 0.16 (6.68–7.18)
Pteronotus quadridens	M	5	14.72 ± 0.15 (14.51–14.88)	13.61 ± 0.16 (13.44–13.78)	5.90 ± 0.04 (5.84–5.95)	4.63 ± 0.11 (4.55–4.70)
	F	9	14.58 ± 0.25 (14.18–14.88)	13.41 ± 0.21 (13.16–13.77)	5.85 ± 0.11 (5.67–6.01)	4.63 ± 0.07 (4.49–4.79)
Stenoderma rufum	M	15	22.15 ± 0.81 (21.37–22.72)	18.51 ± 0.76 (17.82–19.47)	6.67 ± 0.38 (6.43–7.15)	5.79 ± 0.32 (5.54–6.15)
	F	15	23.04 ± 0.61 (22.56–23.63)	19.37 ± 0.71 (18.84–20.21)	6.94 ± 0.27 (6.72–7.16)	6.04 ± 0.30 (5.76–6.36)
Tadarida brasiliensis	M	11	15.97 ± 0.25 (15.62–16.33)	14.67 ± 0.18 (14.24–14.88)	5.57 ± 0.08 (5.40–5.68)	4.33 ± 0.09 (4.16–4.49)
	F	0				

[1]Cranial measurements (mean ± standard deviations in parentheses) of male and female bats from Puerto R
All specimens were from the collection at the Texas Tech Museum. All measurements are in millimeters.
[2]n=1

Zygomatic width	Width of postorbital constriction	Width of braincase	Width across upper molars	Width across upper canines	Greatest length of mandible
11.02 ± 0.16	4.00 ± 0.09	9.65 ± 0.16	7.37 ± 0.09	5.34 ± 0.12	14.58 ± 0.17
(10.70–11.37)	(3.85–4.14)	(9.48–10.13)	(7.20–7.61)	(5.08–5.52)	(14.22–14.90)
10.85 ± 0.16	4.00 ± 0.17	9.50 ± 0.10	7.35 ± 0.09	5.27 ± 0.11	14.39 ± 0.22
(10.59–11.17)	(3.82–4.41)	(9.33–9.69)	(7.17–7.48)	(5.05–5.50)	(14.11–14.93)
7.55 ± 0.03	3.08 ± 0.11	6.75 ± 0.09	5.39 ± 0.10	4.19 ± 0.08	10.22 ± 0.13
(7.52–7.59)	(2.91–3.22)	(6.61–5.22)	(5.22–5.46)	(4.10–4.31)	(10.06–10.38)
7.51 ± 0.16	3.11 ± 0.10	6.80 ± 0.18	5.29 ± 0.11	4.03 ± 0.16	10.11 ± 0.19
(7.31–7.74)	(2.97–3.23)	(6.57–7.17)	(5.11–5.45)	(3.78–4.21)	(9.77–10.43)
14.68 ± 0.46	5.83 ± 0.35	10.84 ± 0.46	9.39 ± 0.27	5.77 ± 0.35	12.17 ± 0.55
(14.38–15.16)	(5.04–5.71)	(10.41–11.27)	(9.13–9.61)	(5.41–6.24)	(11.70–12.81)
15.24 ± 0.60	5.67 ± 0.30	11.00 ± 0.60	9.75 ± 0.39	5.91 ± 0.24	12.87 ± 0.41
(14.78–15.75)	(5.36–5.91)	(10.37–11.54)	(9.32–10.02)	(5.72–6.09)	(12.45–13.23)
9.27 ± 0.16	3.83 ± 0.09	7.93 ± 0.09	6.63 ± 0.09	3.96 ± 0.07	10.48 ± 0.13
(9.07–9.49)	(3.67–4.04)	(7.77–8.07)	(6.49–6.74)	(3.86–4.07)	(10.24–10.62)

Appendix 4
Ectoparasites from Bats on Puerto Rico

PARASITE	Artibeus jamaicensis	Brachyphylla cavernarum	Eptesicus fuscus	Erophylla sezekorni	Lasiurus borealis	Molossus molossus
ACARINA						
Argasidae						
Ornithodoros viguerasi				X		
Ornithodoros sp.				X		
Labidocarpidae						
Lawrenceocarpus micropilus		X				
Lawrenceocarpus puertoricensis		X				
Paralabidocarpus artibei	X					
Paralabidocarpus foxi	X					
Paralabidocarpus stenodermi						
Macronyssidae						
Radfordiella oudemansi		X				
Spelaeorhynchidae						
Spelaeorynchus monophylli						
Spelaeorynchus praecursor	X					
Spinturnicidae						
Cameronieta thomasi						
Periglishcurus cubanus		X		X		
Periglischrus iheringi	X		X	X		
Periglischrus vargasi	X					
Spinturnix bakeri			X			
INSECTA						
Streblidae						
Aspidoptera phyllostomatus	X					
Megistopoda aranea	X					
Nycterophilia parnelli						
Paradyschria fusca						
Trichobius cernyi						
Trichobius intermedius	X					
Trichobius robynae	X			X		
Trichobius sp.						
Trichobius runcatus		X		X		

Sources: 1 = Anthony, 1918; 2 = Fain et al., 1967; 3 = Gannon and Willig, 1994b; 4 = Gannon and Willig, 1995; 5 = Rudnick, 1960; 6 = Tamsitt and Fox, 1970a; 7 = Tamsitt and Fox, 1970b; 8 = Tamsitt and Valdivieso, 1970; and 9 = Webb and Loomis, 1977.

Monophyllus redmani	Mormoops blainvillii	Noctilio leporinus	Pteronotus parnellii	Pteronotus quadridens	Stenoderma rufum	Tadarida brasiliensis	Source
							6
							3, 4
							7, 9
							9
					X		7, 9
					X		9
					X		9
							6, 9
X							2, 3, 6, 8, 9
X							2, 3, 4, 6, 9
			X				3
							3
X					X		3, 4, 5, 6, 8, 9
X							3, 8, 9
							3
							3, 4
							3, 4, 9
X							3, 4
		X					1
X							3
X							3, 4, 9
X							9
X							3
X							3, 6, 9

Appendix 5

Chromosomal Characteristics

Species	Diploid number	Fundamental number	Source
Artibeus jamaicensis	30/31[1]	60	Baker and Lopez, 1970
Brachyphylla cavernarum	32	60	Baker and Lopez, 1970
Eptesicus fuscus	50	48	Baker and Lopez, 1970
Erophylla sezekorni	32	60	Baker and Lopez, 1970
Lasiurus borealis	28	48	Baker and Patton, 1967
Molossus molossus	48	56	Baker and Lopez, 1970
Monophyllus redmani	32	60	Baker and Lopez, 1970
Mormoops blainvillii	38	60	Nagorsen and Peterson, 1975
Noctilio leporinus	34	62	Baker, 1970
Pteronotus parnellii	38	60	Baker and Lopez, 1970
Pteronotus quadridens	38	60	Baker and Lopez, 1970
Stenoderma rufum	30/31*	56	Baker and Lopez, 1970

[1]In *Artibeus* and *Stenoderma,* the X chromosome can be translocated to autosomes, but this translocation does not occur in all individuals (Wetterer et al., 2000).

Appendix 6
Technical Names of Plants Mentioned in the Text

Achiote	*Bixa orellana*
Angelin	*Andira inermis*
Bamboo	*Gigantochloa scoruchinii*
Banana	*Musa paradisiaca*
Bromeliad	Bromeliaceae
Bullet-wood	*Manilkara bidentata*
Cactus	Cactaceae
Cassava	*Manihot esalenta*
Century plant	*Agave*
Chicken tree	*Pterocarpus officinalis*
Columnar cactus	*Pilosocereus* and *Stenocereus*
Corn	*Zea mays*
Elder	*Piper aduncum, P. glabra, P. hispidum*
Fern	Pterophyta
Fig	*Ficus*
Ginger	*Zingiber officinale*
Golden apple	*Spondias dulsis*
Grape	*Vitis vinifera*
Guanábana	*Annona muricata*
Guava	*Psidium guajava*
Heliconia	Heliconiaceae
Hickory	*Carya*
Hogplum	*Spondias mombin*
Jag	*Ficus citrifolia*
Jobo	*Spondias mombin*
Liverwort	Hepatophyta
Maga	*Maga grandiflora*
Mallow	Malvaceae
Mamey	*Mammea americana*
Mango	*Mangifera indica*
Mangrove	*Avicennia germinans, Conocarpus erectus, Laguncularia racemosa, Rhizophora mangle*
María	*Calophyllum calaba*
Mountain immortelle	*Erythrina poeppigiana*
Oak	*Quercus*
Orchid	Orchidaceae
Palm	*Palmae*
Panama berry	*Mutingia calabura*
Papaya	*Carica papaya*
Philodendron	*Philodendron*

Pineapple	*Ananas comosus*
Pitch apple	*Clusea rosea*
Portia	*Thespesia populnea*
Rod-wood	*Eugenia jambos*
Royal palm	*Roystonea*
Sapodilla	*Manilkara bidentata, M. zapota*
Sierra palm	*Prestoea montana*
Silk cotton	*Ceiba petandra*
Sugarcane	*Saccharum officinarum*
Sweetgum	*Liquidamber stryaciflua*
Sweet potato	*Ipomoea batatas*
Tobacco	*Nicotiana tabacum*
Trumpet tree	*Cecropia schreberiana*
Tulip tree	*Liriodendron tulipifera*
Turkey berry	*Solanum torvum*
White siris	*Albizia procera*
Wild tamarind	*Leucaena leucocephala*
Woman's tongue	*Albizia lebbek*
Yam	*Ipomoea batatas*
Yautía	*Xanthosoma sagittifolium*

Appendix 7
Technical Names of Animals Mentioned in the Text[1]

American kestrel	*Falco sparverius*
Ant	Formicidae
Antillean ghost-faced bat	*Mormoops blainvillii*
Antillean fruit bat	*Brachyphylla cavernarum*
Bald eagle	*Haliaeetus leucocephalus*
Ballyhoo	*Hemiramphs brasiliensis*
Bamboo bat	*Tylonycteris*
Bat falcon	*Falco rufigularis*
Batbug	Cimicidae
Bear	*Ursus*
Beetle	Coleoptera
Big brown bat	*Eptesicus fuscus*
Black rat (house rat)	*Rattus rattus*
Brazilian free-tailed bat	*Tadarida brasiliensis*
Brown flower bat	*Erophylla sezekorni*
Chigger	Trombiculidae
Cicada	Cicadidae
Common long-tongued bat (Pallas's long-tongued bat)	*Glossophaga soricina*
Common pipistrelle	*Pipistrellus pipistrellus*
Common vampire bat (vampire bat)	*Desmodus rotundus*
Coquí	*Eleutherodactylus coqui*
Corn earworm	*Helicoverpa zea*
Corn rootworm	*Diabrotica*
Cricket	Orthoptera
Damselfly	Odonata
Davy's naked-backed bat	*Pteronotus davyi*
Deer mouse	*Peromyscus maniculatus*
Dragonfly	Odonata
Dwarf dog-faced bat	*Molossops temminckii*
Elephant	Elephantidae
False vampire bat (spectral bat)	*Vampyrum spectrum*
Flying lemur	Dermoptera
Flying squirrel	Petromyinae
Ghost bats	*Diclidurus*
Grasshopper	Orthoptera
Great evening bat	*Ia io*
Great noctule	*Nyctalus lasiopterus*
Greater Antillean long-tongued bat (Leach's single leaf bat)	*Monophyllus redmani*

Greater bulldog bat	*Noctilio leporinus*
Greater spear-nosed bat	*Phyllostomus hastatus*
Ground squirrel	*Spermophilus*
Guinea pig	*Cavia porcellus*
Hairy-legged vampire bat	*Diameus youngii*
Hoary bat	*Lasiurus cinereus*
Hog-nosed bat	*Craseonycteris thonglongyai*
House mouse	*Mus musculus*
Jamaican fruit bat (Jamaican fruit-eating bat)	*Artibeus jamaicensis*
Key West quail-dove	*Geotrygon chysia*
Lacewing	Neuroptera
Leafhoppers	Homoptera
Lesser bamboo bat	*Tylonycteris pachypus*
Lesser bulldog bat	*Noctilio albiventris*
Little brown bat	*Myotis lucifugus*
Long-eared myotis	*Myotis evotis*
Malayan flying fox (large flying fox)	*Pteropus vampyrus*
Mato Grosso dog-faced bat	*Molossops mattogrossensis*
Merlin	*Falco columbarius*
Mexican fish-eating bat (fish-eating bat)	*Myotis vivesi*
Mongoose	*Herpestes javanicus*
Moth	Lepidoptera
Hairless bat	*Cheiromeles torquatus*
New Zealand lesser short-tailed bat	*Mystacina tuberculata*
Norway rat (brown rat)	*Rattus norvegicus*
Old World fruit bats	Pteropodidae
Osprey	*Pandion haliaetus*
Parnell's mustached bat	*Pteronotus parnellii*
Peter's ghost-faced bat (ghost-faced bat)	*Mormoops megalophylla*
Puerto Rican boa	*Epicrates inornatus*
Puerto Rican parrot	*Amazona vittata*
Puerto Rican short-eared owl	*Asio flammeus*
Puerto Rican vine snake	*Alsophis portoricensis*
Puerto Rican whip-poor-will	*Caprimulgus noctitherus*
Rabbit	Leporidae
Red bat	*Lasiurus borealis*
Red fig-eating bat (red fruit bat)	*Stenoderma rufum*
Red-tailed hawk	*Buteo jamaicensis*
Roach	Dictyoptera

Roberts's flatheaded bat	*Mormopterus petrophilus*
Roundworm	Nematoda
Seminole bat	*Lasiurus seminolus*
Shrew	Soricidae
Silver-haired bat	*Lasionycteris noctivagans*
Silversides	*Atherinomorus stipes*
Sooty mustached bat	*Pteronotus quadridens*
Southern yellow bat	*Lasiurus ega*
Spiny-headed worm	Acanthocephala
Spotted bat	*Euderma maculatum*
Swallow	Hirundinidae
Swift	Apodidae
Tapeworm	Cestoda
Tent-making bat	*Uroderma bilobatum*
Termite	Isoptera
Thrip	Thysanoptera
Tilapia	*Oreochromis mossambicus*
True bug	Diptera
Velvety free-tailed bat (Pallas's mastiff bat)	*Molossus molossus*
Whale	Cetacea
White-lined bat (broad-nosed bat)	*Platyrhinus*
White-winged vampire bat	*Diphylla ecuadata*
Wooly bat	*Kerivoula*

[1]Common names of mammals that were suggested by Wilson and Cole (2002) are given in parentheses when they differ from common names used in this book.

Localities where bats have been recorded in the Commonwealth of Puerto Rico. Row numbers correspond to localities shown on range maps contained within each species account and on map 18. Column numbers indicate source of the information.

Sources are: 1 = American Museum of Natural History; 2 = University of California at Berkeley, Museum of Vertebrate Zoology; 3 = Carnegie Museum of Natural History; 4 = Field Museum of Natural History; 5 = University of Kansas, Museum of Natural History; 6 = Timm and Genoways (2003) and University of Kansas, Museum of Natural History; 7 = Los Angeles County Museum; 8 = Louisiana State University, Museum of Zoology; 9 = Harvard University, Museum of Comparative Zoology; 10 = University of Oklahoma, Sam Noble Oklahoma Museum of Natural History; 11 = Royal Ontario Museum; 12 = Texas A&M University Mammal Collection; 13 = Texas Tech University, Museum; 14 = University of Michigan, Museum of Zoology; 15 = United States National Museum, Smithsonian Institution; 16 = Anthony (1918, 1925); 17 = Brooke (1994); 18 = Buden (1976); 19 = Conde Costas and Gonzalez (1990); 20 = Hall and Tamsitt (1968); 21 = Heatwole et al. (1963); 22 = Heatwole et al. (1964); 23 = Jackson (1916); 24 = Kunz et al. (1983); 25 = Miller (1899); 26 = Miller (1900); 27 = Miller (1913); 28 = Rodriguez and Reagan (1984); 29 = Rodríguez-Durán and Lewis (1987); 30 = Rodríguez-Durán (1998); 31 = Rodríguez-Durán (1999); 32 = Rodríguez-Durán and Vásquez (2001); 33 = Starrett and Rolle (1962); 34 = Tamsitt and Valdivieso (1970); 35 = Thomas and Thomas (1974); 36 = A. Rodríguez-Durán (unpublished observations); and 37 = Gannon (unpublished observations).

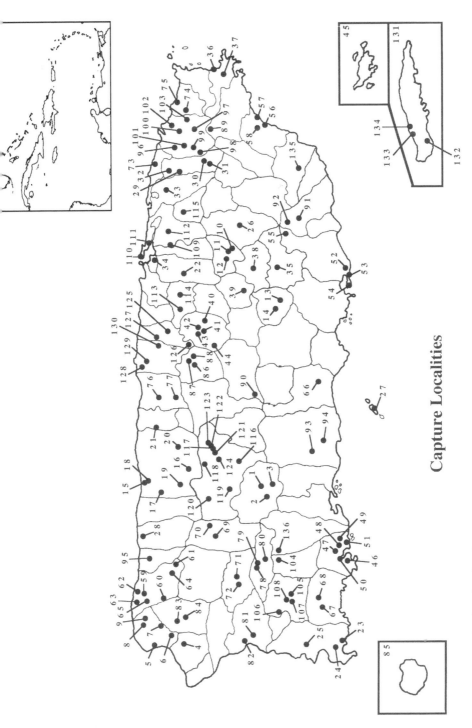

Map 18 Capture localities listed in the gazetteer.

Map 19 Major rivers of Puerto Rico.

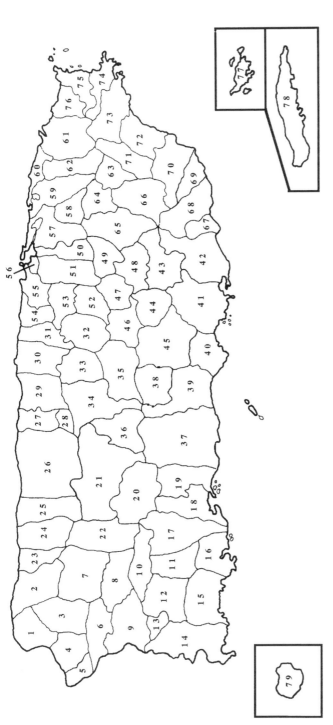

Map 20 Municipalities of Puerto Rico. Numbers correspond as follows: Adjuntas 20; Aguas 4; Aguadilla 1; Aguas Buenas 49; Aibonito 44; Añasco 6; Arecibo 26; Arroyo 67; Barceloneta 27; Barranquitas 46; Bayamón 51; Cabo Rojo 14; Caguas 65; Camuy 24; Canóvanas 62; Carolina 59; Cataño 56; Cayey 43; Ceiba 74; Ciales 34; Cidra 48; Coamo 45; Comerio 47; Corozal 32; Culebra 77; Dorado 54; Fajardo 75; Florida 28; Guánica 16; Guayama 42; Guayanilla 18; Guaynabo 50; Gurabo 64; Hatillo 25; Hormigueros 13; Humacao 72; Isabela 2; Jayuya 36; Juana Díaz 39; Juncos 63; Lajas 15; Lares 22; Las Marías 8; Las Piedras 71; Loíza 60; Luquillo 76; Manatí 29; Maricao 10; Maunabo 69; Mayagüez 9; Moca 3; Mona 79 Morovis 33 Naguabo 73 Naranjito 52; Orocovis 35; Patillas 68; Peñuelas 19; Ponce 37; Quebradillas 23; Rincón 5; Río Grande 61; Sabana Grande 11; Salinas 41; San Germán 12; San Juan 57; San Lorenzo 66; San Sebastián 7; Santa Isabel 40; Toa Alta 53; Toa Baja 55; Trujillo Alto 58; Utuado 21; Vega Alta 31; Vega Baja 30; Vieques 78; Villalba 38; Yabucoa 70; and Yauco 17.

Locality No.	Locality	Artibeus jamaicensis	Brachyphylla cavernarum	Eptesicus fuscus	Erophylla sezekorni	Lasiurus borealis	Molossus molossus
1	Adjuntas: Town of Adjuntas						16
2	Adjuntas: Guilarte State Forest, various locations within the forest	13	13	13	13		13
3	Adjuntas: Guilarte State Forest, 2.5 km on road 388 from intersection with road 518						
4	Aguada: Town of Coloso, Coloso sugar mill, inside the sugar storage building	36					
5	Aguadilla: Town of Aguadilla		11				7
6	Aguadilla: Cucaracha Cave, along road 443, ca. 0.4 km N road 111, Barrio Caimital Bajo						
7	Aguadilla: Madama Cave,Barrio Caimital Bajo, off road 443 ca. 0.4 km S of road #2						
8	Aguadilla: Puente del Cedro roost (abandoned dam), Barrio San Antonio, down a creek that passes under road 466				36		
9	Aguadilla: Tunel Playa roost (storm sewer), Barrio San Antonio, down a creek that passes under road 466			30			
10	Aguas Buenas: 5.6 km SW of Aguas Buenas Cave			6			
11	Aguas Buenas: Aguas Buenas Cave, 5.6 km SW of town of Aguas Buenas,Guayana Prov., 18° 14' 01" N, 66° 06' 30" W	6, 9, 15, 24	11	9, 11, 13		11	
12	Aguas Buenas: Barrio Sumidero, Sector La Capilla, dirt road off road 794						

Monophyllus ...mani	Mormoops blainvillii	Noctilio leporinus	Pteronotus parnellii	Pteronotus quadridens	Stenoderma rufum	Tadarida brasiliensis
						15
13			13		13	
					13	
8						
29	29			29		
30	30			30		
11				11		
13						

Locality No.	Locality	Artibeus jamaicensis	Brachyphylla cavernarum	Eptesicus fuscus	Erophylla sezekorni	Lasiurus borealis	Molossus molossus
13	Aibonito: Town of Aibonito						16
14	Aibonito: San Cristobal Canyon	13					
15	Arecibo: Town of Arecibo	15					15
16	Arecibo: Cueva Matos, road 10, between towns of Arecibo and Utuado	36					
17	Arecibo: Culebrones Cave (aka Mata de Platano Cave), Barrio Dominguito, 7 km SW of town of Arecibo, away from any secondary road	13	13, 28, 30	30	13, 28		
18	Arecibo: Los Chorros Cave near town of Arecibo						
19	Arecibo: Río Abajo State Forest, various locations within the forest	9, 13	13		13		
20	Arecibo: Ventana Cave, 5.1 km south of Utuado exit on Puerto Rico Highway 10	13	13		13		
21	Barceloneta: Cambalache State Forest, various locations within the forest	13	13	13	13		
22	Bayamón: 3.2 km W town of Bayamón, near Limestone cave	9, 14			9, 15, 25		
23	Cabo Rojo: Boquerón State Forest, Manglares de La Pitahaya-La Pargura, 0.65 km E of West Gate section						
24	Cabo Rojo: Boquerón National Wildlife Refuge (USFWS)		13				
25	Cabo Rojo: Town of Cabo Rojo						
26	Caguas: Town of Caguas						15
27	Caja de Muertos: On the island (location not specified)	6, 9					
28	Camuy: Amador Cave, off road 485 in front of Peñón de los Amador beach	30		36	36		

Monophyllus ...dmani	Mormoops blainvillii	Noctilio leporinus	Pteronotus parnellii	Pteronotus quadridens	Stenoderma rufum	Tadarida brasiliensis
13						
13, 28	13, 28			28, 30	36	
	11					
13	13		13	13	13	9
13		36		13	13	
				13	13	
13, 16, 26						
13	13					
13	13	13	13			
						16
						15
		30				

Locality No.	Locality	Artibeus jamaicensis	Brachyphylla cavernarum	Eptesicus fuscus	Erophylla sezekorni	Lasiurus borealis	Molossus molossus
29	Canóvanas: 10 km S town of Canóvanas			9			
30	Canóvanas: Caribbean National Forest, Arroyo Section, 1 km SE of road 907, 2.8 km S, 2.0km W El Verde	13			13		
31	Canóvanas: Caribbean National Forest, La Condesa Section, off road 946, 0.75 km S of Benitez, 3.7 km from road 186, 490m	13	13	13	13		
32	Canóvanas: Town of Canóvanas, in building						21
33	Carolina: 4.0 km NE town of Carolina, Cueva Cerro de San Jose		9				
34	Catano: Buchanan		7				
35	Cayey: various locations in town of Cayey	15	1				11
36	Ceiba: Ceiba State Forest, 2.3 km S, 3.5 km E town of Fajardo, near Sonda de Vieques beach, 1m	13					
37	Ceiba: Roosevelt Roads Naval Base, 1.5 km W of hospital	13					
38	Cidra: 1 km NE town of Cidra	6					
39	Comercio: La Mora Cave, Barrio Piñas, off road 775, near Iglesia de La Mora	15	15				
40	Corozal: 1.5 km E of town Corozal					13	
41	Corozal: 1.6 km W of town Corozal	13	13			13	13
42	Corozal: Los Quintero Cave, Barrio Cibuco	30					
43	Corozal: Convento Cave, 18° 21' N, 66° 18' W	32					
44	Corozal: Mirante Sur, 6.4 km N (= South, typo) from Highway 2 on Highway 160						11, 12
45	Culebra: Various locations on the Island	15, 22					9, 22
46	Guánica: 3 km W, 1 km SE town of Ensenada						

Monophyllus dmani	Mormoops blainvillii	Noctilio leporinus	Pteronotus parnellii	Pteronotus quadridens	Stenoderma rufum	Tadarida brasiliensis
13					13	
13				13	13	
			9			
13						
6						
13					13	
30			30			
		17				
11						

Locality No.	Locality	Artibeus jamaicensis	Brachyphylla cavernarum	Eptesicus fuscus	Erophylla sezekorni	Lasiurus borealis	Molossus molossus
47	Guánica: 3.5 km E, 1 km N town of Guánica						
48	Guánica: 5 km E town of Guánica	8	8				
49	Guánica: 7.5 km E town of Guánica		6				
50	Guánica: Town of Guánica				19		
51	Guánica: Guánica State Forest, various locations within the forest	13, 19	13, 19		13, 19		
52	Guayama: Aguirre State Forest, 1.8 km SW on road 7710 from intersection with road 3, then 0.8 km S on unpaved road	13					
53	Guayama: Aguirre State Forest, 5.9 km SW on road 7710 from intersection with road 3, then 0.5 km S on unpaved road	13					
54	Guayama: Aguirre State Forest, road 7710 SW to end of pavement, then 0.5 km SW						
55	Guayama: Carite State Forest, various locations within the forest	13	13	13			
56	Humacao: 3.2 km S town of Punta Santiago						2
57	Humacao: Playa Humacao						
58	Humacao: Town of Punta Santiago	2	2				2
59	Isabela: 4 km S of Cueva de los Alfaros	8		9			
60	Isabela: Caidas Cave, Barrio Arenales, off road 112 ca. 6 km S of road 2	30	30				
61	Isabela: Casa de Piedras Forest Preserve						
62	Isabela: Cave 1.6 km E Mora near town of Isabela		9				
63	Isabela: Cueva de San Felipe #3	8					
64	Isabela: Guajataca State Forest, various locations within the forest	13, 30	13, 30		11, 13, 30		

Monophyllus edmani	Mormoops blainvillii	Noctilio leporinus	Pteronotus parnellii	Pteronotus quadridens	Stenoderma rufum	Tadarida brasiliensis
8				8		
11				8		
6				6		
13, 15, 19	13, 19		13			
13		13				
13						
13			13			
2						
				2		
					15	
13, 30	13, 30		13, 30	13, 30		

Locality No.	Locality	Artibeus jamaicensis	Brachyphylla cavernarum	Eptesicus fuscus	Erophylla sezekorni	Lasiurus borealis	Molossus molossus
65	Isabela: km 111.4 between towns of Aguadilla and Arecibo, 4 km S town of Isabela	9					
66	Juana Díaz: Fort Allen Naval Base, various locations on the base	13			13		
67	Lajas: Town of Monte Grande, Candelaria, Lower Cordillera						
68	Lajas: Nombres Cave, Parguera, dirt road off road 117	30					
69	Lares: Cundo Reyes Cave, Barrio Piletas Soyer, dirt road off road 453	30					
70	Lares: Cavernas del Río Camuy Park, various locations in the park	13, 30	13		13		
71	Las Marías: Espinosas y Barrientos Cave, dirt road off road 120 at km marker 31.4		30				
72	Las Marías: Town of Las Marías	7, 15					
73	Loíza: Town of Old Loíza, Piedra de la Cueva	9, 14					
74	Luquillo: Bisley Trail off Rt 998, 280 m, in the Caribbean National Forest						
75	Luquillo: Town of Luquillo						
76	Manatí: Town of Manatí						
77	Manatí: Río Encantado Cave, dirt road off road 642	30					
78	Maricao: Town of Maricao			15, 23			
79	Maricao: Maricao State Forest, 1.0 km from town of Maricao near Fish Hatchery	13, 15			13		
80	Maricao: Maricao State Forest, various locations within the forest	13	13		13		
81	Mayagüez: 8.1 km SE town of Mayagüez	6		6			
82	Mayagüez: Town of Mayagüez	9					11, 15

Monophyllus redmani	Mormoops blainvillii	Noctilio leporinus	Pteronotus parnellii	Pteronotus quadridens	Stenoderma rufum	Tadarida brasiliensis
9			9			
13				13		
				15		14
30		30				
13, 30	13, 30		13, 30			
30						
		14, 15				
					13	
15, 27						
						14
15	13				13	
13	13				13	
	15			9, 15		15

Locality No.	Locality	Artibeus jamaicensis	Brachyphylla cavernarum	Eptesicus fuscus	Erophylla sezekorni	Lasiurus borealis	Molossus molossus
83	Moca: Golondrinas Cave I, Barrio Cuchillas tercer piso, Loperena sector, dirt road off road 4444	30					
84	Moca: La Mina Cave, Barrio Cuchillas secondary dirt road off 4444, 18° 24' N, 67° 07'W	30			7, 30	8, 33	
85	Mona: On the island (location not specified)						
86	Morovis: 3.2 km W town of Morovis, in cave	9					
87	Morovis: Cueva Catedral near town of Morovis		1				
88	Morovis: Town of Morovis	9					
89	Naguabo: 0.8 km on Mt. Britton access road, km 13.5 on road 191, Caribbean National Forest	13					
90	Orocovis: Toro Negro State Forest, various locations within the forest	13	13	13	15		
91	Patillas: Carite State Forest, 3 km N, 2.5 km W Compartemento Real						
92	Patillas: Carite State Forest, various locations within the forest	13	13	13	13		
93	Ponce: Hacienda Buena Vista, 16.8 km N the City of Ponce on road 10, at northern end of the park, 200 m	13	13				13
94	Ponce: City of Ponce						
95	Quebradillas: Tunel Negro (abandoned railroad tunnel), Barrio Terra Nova, ca. 0.1 km south of road #2, east of Guajataca River				13		
96	Río Grande: 3.2 km SSE town of El Verde				1, 18		
97	Río Grande: Caribbean National Forest, Base of El Toro Trail off road 191, 750 M, 18° 17' 59" N, 65° 47' 35" W	3, 10, 13	10, 13	13	10, 11, 13		

Monophyllus redmani	Mormoops blainvillii	Noctilio leporinus	Pteronotus parnellii	Pteronotus quadridens	Stenoderma rufum	Tadarida brasiliensis
30	30					
	15, 16	16				
						9, 14
1	14				15	
13						
13			13	11	13	
8			8			
13	13				13	
			13			
			8			9
						13, 30
3, 10, 13			13	13		

Locality No.	Locality	Artibeus jamaicensis	Brachyphylla cavernarum	Eptesicus fuscus	Erophylla sezekorni	Lasiurus borealis	Molossus molossus
98	Río Grande: Caribbean National Forest, El Toro Trail, 9.12 km SW of El Verde Field Station on road 186, 0.5 km up trail		13		13		
99	Río Grande: Caribbean National Forest, Forest Service road 911, 1.12 km SE of road 186			13			
100	Río Grande: Catalina Field Office (USFS), Intersection of Road 191 and 988 in the Caribbean National Forest, 18° 20' 23" N, 65° 45' 44" W	16					
101	Río Grande: El Verde Field Station, Caribbean National Forest, near road 186, 350 m, 18° 19' 18" N, 65° 49' 12" W	3, 13, 34	3, 13	13	13, 34	13	
102	Río Grande: Town of Mameyes						15
103	Río Grande: Santa Catalina, W fork of Río Mameyes	15					
104	Sabana Grande: 1 km E, 4 km N Susúa State Forest, across riverbed of Río Loco	8, 30	8, 30				
105	San Germán: Abejas Cave, close to Murciélagos Cave, Barrio Hoconuco Bajo, dirt road off 358	30					
106	San Germán: Del Viento Cave System, Barrio Duey Alto road 330, ca. 1 km from road 2 across Duey River Bridge	8		8			
107	San Germán: Monte Grande, 4.5 km SE of town of Cabo Rojo	11, 16	9	16			14
108	San Germán: Murciélagos Cave, Barrio Hoconuco Bajo, dirt road off 358		30				
109	San Juan: Caneja Cave, off the intersection of De Diego Expressway and Kennedy Ave.	30					
110	San Juan: City of Old San Juan	14	1, 9, 14	16	15, 16, 18		4
111	San Juan: Isla Verde Airport						11
112	San Juan: Town of Río Piedras	7	14			31	

Monophyllus redmani	Mormoops blainvillii	Noctilio leporinus	Pteronotus parnellii	Pteronotus quadridens	Stenoderma rufum	Tadarida brasiliensis
				13		
13						
13, 34			13	13	1, 3, 5, 11, 13, 20, 34	
					8	
30						
8	8					
		11		9		9, 16
30						
1			15	4, 14		
		9				

Locality No.	Locality	Artibeus jamaicensis	Brachyphylla cavernarum	Eptesicus fuscus	Erophylla sezekorni	Lasiurus borealis	Molossus molossus
113	Toa Alta: Bonita Cave, Barrio Río Lajas, Marzán, next to the intersection of roads 820 and 823	30	30		30		
114	Toa Alta: Town of Toa Alta	16					
115	Trujilo Alto: Cueva de Trujilo Alto	16	1				
116	Utuado: 13.5 km E, 1.5 km N town of Adjuntas		8				
117	Utuado: 8.9 km NE town of Utuado				5		
118	Utuado: Barrio Río Arriba, 9.65 km NNE town of Utuado		9				
119	Utuado: Barrio Santa Rosa	9					
120	Utuado: Camuy Cave						
121	Utuado: Hacienda Rosas, 10 km NE town of Utuado	15			15		
122	Utuado: Hacienda Rosas, 17.5 km NE of town of Utuado 340 m						
123	Utuado: Hacienda Rosas, 17.7 km NE of Utuado 340 m		6				
124	Utuado: Town of Utuado	9					16
125	Vega Alta: Del Humo Cave, Barrio Maricao adentro upstream Mavilla river from road 677	30					
126	Vega Alta: Town of Candelaria						15
127	Vega Alta: Vega State Forest, various locations within the forest			13	13		
128	Vega Baja: Golondrinas Cave II, Tortugero State Forest	30	30				
129	Vega Baja: Near town of Vega Baja	9, 14					
130	Vega Baja: Vega State Forest, various locations within the forest	13			13		
131	Vieques: On the island (location not specified)	15					15
132	Vieques: Vieques National Wildlife Refuge, various locations within the refuge	37					

Monophyllus redmani	Mormoops blainvillii	Noctilio leporinus	Pteronotus parnellii	Pteronotus quadridens	Stenoderma rufum	Tadarida brasiliensis
30	30	30	30			
1	16		16	16		
9						
-						6
			30			
15					15	
					8, 35	
6					8	
						16
30	30		30	30		
					13	
		37			37	

Locality No.	Locality	Artibeus jamaicensis	Brachyphylla cavernarum	Eptesicus fuscus	Erophylla sezekorni	Lasiurus borealis	Molossus molossus
133	Vieques: Abandoned house on route 200, 180 08' 23.0" N, 650 28' 49.4" W						37
134	Vieques: Abandoned house on route 200, 180 07' 36.3" N, 650 31' 20.7" W	37					
135	Yabucoa: Town of Yabucoa						11
136	Yauco: Susúa State Forest, various locations within the forest	13	13		13	13	

Localities Not Plotted—The following localities could not be plotted with the information provided by the source

	Locality	Artibeus jamaicensis	Brachyphylla cavernarum	Eptesicus fuscus	Erophylla sezekorni	Lasiurus borealis	Molossus molossus
	Arecibo: Cueva de Los Indios, near Cueva de Burros, Juncos Central			15			
	Corozal: Cueva de Corozal, San Juan Province	9	1, 9, 11		11		
	Lares: La Cueva Pajita	9, 15					
	Quebradillas: Cueva de las Golondrinas	15					
	Trujilo Alto: La Cueva de Mollfulleda		15				
	Vega Baja: Cueva de Burros, between Vega Alta and Vega Baja				1, 9, 18		

Monophyllus redmani	Mormoops blainvillii	Noctilio leporinus	Pteronotus parnellii	Pteronotus quadridens	Stenoderma rufum	Tadarida brasiliensis
13		13		13	13	
						15
11						
15						
				14		

Glossary

Antitragus A fleshy projection associated with the base of the outer (rear) margin of the pinna.

Argasid A "soft tick" in the family Argasidae; adults feed quickly and do not stay attached to the host, although larvae remain on the host for prolonged periods.

Aspect ratio Wingspan divided by width of the wing, which is the distance from the front to the back of the wing, from the thumb to the tip of the fifth finger in a bat.

Autosome A chromosome other than the sex chromosomes.

Basal metabolic rate Lowest rate of energy consumption possible in a bird or mammal without lowering body temperature.

Batbug A wingless, dorsoventrally flattened insect (order Hemiptera, family Cimicidae) that is an external parasite of bats.

Beetle A hard-bodied insect in the order Coleoptera; front wings are hardened into protective covers with the membranous hind wings folded beneath.

Braincase Posterior part of the skull that houses the major portion of the brain.

Breech birth A birth in which the posterior part of a newborn leaves the vagina first.

Bug, true An insect in the order Hemiptera with piercing/sucking mouthparts; anterior of front pair of wings is hardened but posterior of front wings is membranous.

Bulldog bat A bat in the family Noctilionidae.

Calcar A cartilaginous rod extending from the ankle of a bat and supporting the trailing edge of the tail membrane.

Canine Pointed tooth used for piercing food; it is located behind the incisors and is the first tooth in the maxillary bone of a bat.

Cheek teeth Teeth behind the canines; the molars and premolars.

Chigger Ectoparasitic larval stage of a mite in the family Trombiculidae.

Cloaca In nonmammalian vertebrates, it is the common external opening of the digestive, urinary, and reproductive systems.

Centimeter A unit of distance equal to about 0.4 inch; there are about 2.5 centimeters in 1 inch.

Commensal Living in close association with humans.

Condylobasal length Distance from the anterior edge of the premaxillary bones to the posteriormost part of the occipital condyles.

Constant frequency (CF) Adjective describing a sound that does not change in frequency, referring to echolocation calls of bats (Fig. 10).

Convergent evolution Independent development, over evolutionary time, of similar traits in organisms that are not closely related.

Crepuscular Active near dawn and dusk.

Cranium Part of the skull surrounding the brain.

Cusp Point or bump on the chewing surface of a tooth.

Dayroost A place where a bat rests during the day.

Delayed fertilization A reproductive process in many bats. Fertilization (joining of sperm and egg) does not occur until weeks or months after copulation; meanwhile, sperm are stored in the uterus.

Delayed implantation A reproductive process in a number of mammals, including some bats. After fertilization the embryo remains in a quiescent state for weeks or months, before implanting in the uterus and resuming growth.

Dental formula A way of describing the number and kind of teeth (incisors, canines, premolars, and molars) that typify a species. Number of teeth of each type is given by a fraction, with the numerator indicating the number of teeth in one-half the upper jaw, and the denominator indicating the number of teeth in one-half the lower jaw. Right and left sides are assumed equivalent in number and type of teeth present.

Detritivore An animal that eats detritus (decaying organic particles).

Dictyopteran A roach or mantis; an insect in the order Dictyoptera.

Diploid number The number of pairs of chromosomes that is characteristic of a species.

Distal Away from the base or toward the end; opposite of proximal.

Dorsal Refers to the top or back of an animal; opposite of ventral.

Dorsum The top or back of an animal.

Ear, height of Distance from lowest notch at the base of the ear to the highest point of the pinna, excluding any hairs.

Earwig An insect with well-developed pincers at the rear of the abdomen; order Dermaptera.

Echolocation Process by which bats locate and identify objects by emitting high-frequency sounds and interpreting the returning echoes.

Ectoparasite A parasite that lives on the outer surface of its host.

Endemic Restricted to a particular geographic area or region.

Endoparasite A parasite that lives inside the tissues or body cavities of its host.

Epiphyte A plant that derives moisture and nutrients from the air rather than through a system of roots; an epiphyte often grows on a tree or other plant.

Epithet The part of a scientific name designating the species or subspecies; e.g., in *Homo sapiens,* the word *sapiens* is the specific epithet.

Estrus Time or physiological state when a female is receptive to mating. *See also* postpartum estrus.

Flatworm An elongate, flattened animal lacking legs and a body cavity and placed in the phylum Platyhelminthes; many (flukes and tapeworms) are endoparasites.

Flea Wingless, laterally compressed insect, usually capable of jumping; order Siphonaptera.

Fluke A flatworm in the class Monogenea or Trematoda; trematodes are endoparasites.

Fly An insect with only one pair of wings that is placed in the order Diptera.

Foramen (plural = foramina) A passage through a bone, usually for nerves or blood vessels.

Forearm, length of Distance from elbow to wrist, measured on the folded wing of a bat.

Frequency modulated (FM) Adjective describing a sound that changes continually in frequency, referring to the structure of echolocation calls of bats (Fig. 10).

Free-tailed bat A bat in the family Molossidae.

Fundamental Lowest frequency in a complex sound that contains harmonics.

Fundamental number Number of chromosomal arms in an animal's karyotype.

Fungivore An animal that eats primarily fungi.

g Abbreviation for gram.

Gestation Period of time between fertilization of an egg and birth.

Gram (g) A unit of weight equal to about 0.04 ounce; there are about 28 grams in 1 ounce.

Greatest length of skull Distance from anteriormost tooth or bone to posteriormost projection of skull.

Gular gland External gland located in the throat area of an animal.

Harem In general, a group of females over which a single male has control for mating purposes.

Harmonic Different frequencies in a complex sound that are integral multiples of a base or fundamental frequency.

Harp trap An instrument used to capture flying bats without harm. It usually consists of two sets of vertical wires, separated by 8–10 cm; bats pass through the first set, are stopped by the second set, and fall into a canvas bag that is hung beneath the trap.

Heat *See* estrus.

Hectare A unit of area equaling 10,000 m^2 or about 2.5 acres.

Hibernation A prolonged period of low body temperature and metabolic rate during the cold season of the year.

Hind foot, length of Distance measured from the heel to the tip of the claw on the longest toe.

Histoplasmosis Respiratory disease of mammals caused by inhaling spores of the fungus *Histoplasma capsulatum*.

Holotype Single specimen upon which the name and defining characteristics of a taxon (usually species or subspecies) is based; often synonymous with type.

Home range Area in which an animal carries out its daily activities.

Hot cave A cave with a small entrance that is low relative to most of the underground chambers, resulting in warm air being trapped and an air temperature that is significantly greater than external air temperature.

Hyperthermia A state of elevated body temperature.

Incisor Anteriormost tooth of a bat; any tooth anterior to the canine.

Incisive foramina Paired openings (foramina) through the front of the palate, immediately behind the incisors.

Interfemoral membrane Tail membrane of a bat; synonymous with uropatagium.

Insectivore Depending on context, the term means either a mammal in the order Insectivora, such as shrews and moles, or any kind of animal that eats insects, as many bats do.

Insectivorous Insect-eating.

IUCN International Union for the Conservation of Nature and Natural Resources; an organization with headquarters in Switzerland that promotes maintenance of the diversity of life and ecologically sustainable use of resources.

Karyotype Number and kinds of chromosomes present in an organism.

Keystone taxa Species or other groups of animals that strongly affect structure of a community through their ecological roles.

Kilohertz A unit of frequency, such as the frequency of sound or radiowaves; it is equal to 1,000 cycles per second.

Kilometer A unit of distance equal to about 0.6 mile.

Labidocarpid An ectoparasitic mite belonging to the family Labidocarpidae.

Labio-nasal A compound adjective referring to a structure that is a combination of the lips (labial) and the nose (nasal).

Lacewing A soft-bodied insect in the order Neuroptera with an aquatic larval stage; wings are longer than the cylindrical body, and antennae are long and segmented.

Larynx Cartilaginous box that contains the vocal cords; the voice box.

Lasiurine A bat in the genus *Lasiurus*.

Leafhopper An insect in the order Homoptera that feeds on plant juices using piercing/sucking mouthparts; wings are held tentlike over the body when at rest.

Leaf-nosed bat A bat in the family Phyllostomidae.

Least interorbital width Shortest distance between the orbits measured across the top of the skull.

Macronyssid A medium-sized mite in the family Macronyssidae; often seen crawling on the fur or wings of bats.

Mandible The lower jaw.

Maxillary toothrow Teeth in the maxillary bone, which are the canines and all cheek teeth.

Mayfly A soft-bodied insect with an aquatic larval stage belonging to the order Ephemeroptera; adults have two pairs of membranous wings that are held vertically when at rest.

Metacarpals Bones of the hand that extend from the wrist to the base of the fingers (phalanges).

Metabolic rate Amount of energy used by an animal per unit time.

Meter A unit of distance equal to about 3.3 feet.

Micron A unit of distance equal to about 0.00004 inch; there are 1,000 microns in one millimeter.

Millimeter (mm) A unit of distance equal to about 0.04 inch; there are 10 mm in one centimeter and 1,000 mm in one meter.

Mist net A very fine net made of nylon that is used to capture flying birds or bats.

mm Abbreviation for millimeter.

Molar A posterior cheek tooth, usually with multiple roots. A mammal has only one set of molars in its life; there are no molars among the deciduous (milk) teeth.

Molossid A bat in the family Molossidae; a free-tailed bat.

Monadnock A hill or mountain of erosion-resistant rock atop a peneplane.

Monestrous Coming into heat (estrus) only once each year.

Mormoopid A bat in the family Mormoopidae; a ghost-faced or mustached bat.

Moth A nocturnal insect in the order Lepidoptera with large, broad wings; wings, body, and legs are covered with scales.

Mycelial Referring to the filamentous growth form of a fungus.

Myobiid Small, elongate mites in the family Myobiidae that have front legs modified for grasping hairs of their host.

Neonate A newborn animal.

Nightroost Location where a bat rests during the night.

Noctilionid A bat in the family Noctilionidae; a bulldog bat.

Nocturnal Active at night.

Nose-leaf Fleshy appendage in some bats, such as members of the Phyllostomidae, that arises from the snout near the nostrils.

Nycteribiid Wingless flies with a spiderlike appearance in the family Nycteribiidae; all are ectoparasitic on bats.

Occipital condyle Rounded area at base of skull for articulation with the first vertebra.

Occlusal An adjective referring to the chewing surface of a tooth.

Omnivore An animal that eats many types of foods, including plants and animals.

Palate Roof of the mouth.

Parturition Act of giving birth.

Pelage A general term referring to the hairy coat of any mammal.

Peneplane A large area of land shaped by erosion so that it shows little change in elevation.

Phalanges (singular = phalanx) The bones of each finger.

Phyllostomid A bat in the family Phyllostomidae; a leaf-nosed bat.

Pinna (plural = pinnae) External ear.

Polyctenid An ectoparasitic bug (family Polyctenidae) covered with comblike structures that presumably help the animal to grasp the fur of its host.

Polyestrous Coming into heat (estrus) more than once each year.

Polyestry The pattern of coming into heat (estrus) more than once each year.

Postorbital constriction Site of minimum distance across the skull behind the orbits.

Postpartum estrus A period of sexual receptivity just after a female gives birth.

Premaxillary bone Anterior bone in the rostrum of a bat; the incisors are rooted in the premaxillary bone.

Premolar A cheek tooth between the molars and canine.

Promiscuous Having more than one mate in a single season.

Proximal Referring to a part close to the base or center; opposite of distal.

Rabies A viral disease that attacks the central nervous system of mammals.

Raptor Predatory bird, such as a falcon, hawk, or owl.

Retina Layer of light-sensitive neurons inside the eye responsible for sight.

Riparian Associated with streams or rivers.

Roach A scavenging insect with a flattened, leathery body and long, slender antennae; roaches are placed in the order Dictyoptera along with mantids.

Rosensteiniid A mite in the family Rosensteiniidae; although generally not considered true ectoparasites, these mites occasionally are found on bats and frequently are abundant in guano.

Rostrum Part of the skull that is anterior to the orbits.

Roundworm Small, cylindrical animals with a primitive type of body cavity belonging to the phylum Nematoda; they are extremely abundant and often are endoparasites of other animals.

Saprophyte A microorganism that subsists on decaying matter, contributing to decomposition.

Sarcoptid A mite in the family Sarcoptidae; individuals often are ectoparasites associated with hair follicles.

Sclerophyllous Adjective describing leaves that are rough and waxy.

SD Abbreviation for standard deviation, a statistical estimate of variation around the mean.

Serpentine A greenish rock with a mottled appeareance, consisting primarily of hydrous magnesium silicate.

Spelaeorhynchid A ticklike mite from the Neotropics in the family Spelaeorhynchidae.

Spinturnicid A mite in the family Spinturnicidae; they are common ectoparasites of bats that crawl crablike over the body and wings.

Spiny-headed worm An animal in the phylum Acanthocephala; all members are endoparasites living in the intestines of vertebrates.

Standard deviation (SD) A statistical estimate of variation around the mean.

Streblid Flies belonging to the family Streblidae that are ectoparasitic on bats; most have wings and are capable of weak flight, but they typically remain on the host for most or all of their lives.

Sweep Used as a noun or verb referring to the act of changing frequencies in an echolocation pulse.

Systematist A biologist concerned with the evolutionary relationship between different taxa.

Tabonuco forest A forest association in the tropical wet forest life zone that is dominated by candlewood (tabonuco) trees.

Tail, length of Distance from the base of the tail to the tip of the last vertebra.

Tail membrane A flap of skin stretching from one leg to the other and enclosing some or all tail vertebrae (when present) of a bat; synonymous with inter-femoral membrane and uropatagium.

Tapeworm A flattened, ribbonlike, segmented animal in the phylum Platyhelminthes, class Cestoda; they live within the digestive tract of their host.

Taxon (plural = taxa) A group of similar animals; a species, genus, or family is a taxon.

Taxonomist A biologist concerned with classifying organisms into groups (taxa).

Thermoregulation Maintenance of body temperature at constant levels.

Torpor A reversible state of reduced body temperature and metabolic rate in a mammal or bird.

Total length Distance from the nose to the tip of the last tail vertebra.

Tragus A fleshy projection associated with the base of the inner (front) margin of the pinna.

Translocated A translocation occurs when part or all of a chromosome becomes attached to another chromosome.

Tricolored Having three colors; may refer to an individual hair or to the overall appearance of an animal's fur.

Trombiculid A mite in the family Trombiculidae; larvae (chiggers) are ectoparasites on vertebrates, but other stages of the life cycle are free-living.

Type The single specimen upon which the name of a taxon is based; often synonymous with holotype.

Ultrasound Sound having frequencies above 20 kilohertz, which is the approximate upper limit of human hearing.

Uropatagium The tail membrane; synonymous with interfemoral membrane.

Vascularized Possessing a supply of blood vessels.

Venter The belly and chest of an animal; its ventral surface.

Ventral Referring to the lower parts or underside of an animal; opposite of dorsal.

Vertebra (plural = vertebrae) One of the chain of bones forming the "backbone" and tail.

Vespertilionid A member of the plain-nosed bat family (Vespertilionidae).

Vibrissae Stiff hairs on the face of a mammal that function as tactile organs; the "whiskers."

Yeast A form of a fungus that reproduces by budding.

Zygomatic arch A narrow strip of bone on the side of a mammalian skull that extends from the anterior edge of the eye socket to just in front of the external ear opening.

Zygomatic breadth Greatest distance between the outer portions of the zygomatic arches.

References

Aihartza, J. R., U. Goiti, D. Almenar, and I. Garin. 2003. Evidence of piscivory by *Myotis capaccinii* (Bonaparte, 1837) in southern Iberian Peninsula. *Acta Chiropterologica* 5:193–98.

Altenbach, J. S. 1989. Prey capture by the fishing bats *Noctilio leporinus* and *Myotis vivesi. Journal of Mammalogy* 70:421–24.

Altringham, J. D. 1996. *Bats: Biology and Behaviour*. New York: Oxford University Press.

Altringham, J. D., and M. B. Fenton. 2003. Sensory ecology and communication in the Chiroptera. Pp. 90–127 in *Bat Ecology,* ed. T. H. Kunz and M. B. Fenton. Chicago: University of Chicago Press.

Anthony, H. E. 1918. Indigenous land mammals of Porto Rico, living and extinct. *Memoirs of the American Museum of Natural History,* New Series 2:331–435.

———. 1925. *Mammals of Porto Rico, Living and Extinct. Scientific Survey of Porto Rico and the Virgin Islands. Part 1.* New York: New York Academy of Sciences. 9:1–96.

Arita, H. T. 1993. Conservation biology of the cave bats of Mexico. *Journal of Mammalogy* 74:693–702.

Arita, H. T., and J. Ortega. 1998. The Middle American bat fauna: Conservation in the neotropical-nearctic border. Pp. 295–308 in *Bat Biology and Conservation,* ed. T. H. Kunz and P. A. Racey. Washington, D.C.: Smithsonian Institution Press.

Baker, R. J. 1970. Karyotypic trends in bats. Pp. 65–96 in *Biology of Bats,* vol. 1, ed. W. A. Wimsatt. New York: Academic Press.

Baker, R. J., and H. H. Genoways. 1978. Zoogeography of Antillean bats. Pp. 53–97 in *Zoogeography in the Caribbean,* ed. F. G. Gill. Special Publication 13. Academy of Natural Sciences of Philadelphia.

Baker, R. J., and G. Lopez. 1970. Karyotypic studies of the insular populations of bats on Puerto Rico. *Caryologia* 23:465–72.

Baker, R. J., and J. L. Patton. 1967. Karyotypes and karyotypic variation of North American vespertilionid bats. *Journal of Mammalogy* 48:270–86.

Baker, R. J., P. V. August, and A. A. Steuter. 1978. *Erophylla sezekorni. Mammalian Species* 115:1–5.

Band, A. 2000. Bats of the Cayman Islands. *Bats* 18:9–12.

Barbour, R. W., and W. H. Davis. 1969. *Bats of America.* Lexington: University Press of Kentucky.

Barclay, R. M. R. 1994. Constraints on reproduction by flying vertebrates: Energy and calcium. *American Naturalist* 144:1021–31.

———. 1999. Bats are not birds—a cautionary note on using echolocation calls to identify bats: A comment. *Journal of Mammalogy* 80:290–96.

Barclay, R. M. R., and L. D. Harder. 2003. Life histories of bats: Life in the slow lane. Pp. 209–56 in *Bat Ecology,* ed. T. H. Kunz and M. B. Fenton. Chicago: University of Chicago Press.

Barlow, K. E., N. Vaughan, K. E. Jones, A. Rodríguez-Durán, and M. R. Gannon. 2000. Are bats which pollinate and disperse forest plants particularly sensitive to disturbance? A case study of the effects of Hurricane Georges on bats of Puerto Rico. *Bulletin of the British Ecological Society* 31:36–39.

Bateman, G. C., and T. A. Vaughan. 1974. Nightly activities of mormoopid bats. *Journal of Mammalogy* 55:45–65.

Beatty, H. A. 1944. Fauna of St. Croix, V.I. *Journal of Agriculture,* University of Puerto Rico, 28:181–85.

Beck, B. F., M. Fram, and J. R. Carvajal. 1976. The Aguas Buenas Caves, Puerto Rico: Geology, hydrology, and ecology with special reference to the histoplasmosis fungus. *National Speleological Society Bulletin* 38:1–16.

Bernard, K. W., J. Mallonee, J. C. Wright, F. L Reid, S. Makintubee, R. A. Parker, D. M. Dwyer, and W. G. Winkler. 1987. Preexposure immunization with intradermal human diploid cell rabies vaccine. *Journal of the American Medical Association* 257:1059–63.

Best, T. L., and K. G. Castro. 1981. Synopsis of Puerto Rican mammals. *Studies in Natural Sciences* 2:1–12.

Birdsey, R. A., and P. L. Weaver. 1982. *The Forest Resources of Puerto Rico.* Resource Bulletin SO-85. U.S. Department of Agriculture, Forest Service, Southern Forest Experiment Station. 59 pp.

————. 1987. *Forest Area Trends in Puerto Rico.* Research Note SO-331. U.S. Department of Agriculture, Forest Service, Southern Forest Experiment Station. 5 pp.

Bloedel, P. 1955. Hunting methods of fish-eating bats, particularly *Noctilio leporinus. Journal of Mammalogy* 36:390–99.

Bogdanowicz, W., M. B. Fenton, and K. Daleszczyk. 1999. The relationship between echolocation calls, morphology, and diet in insectivorous bats. *Journal of Zoology* (London) 247:381–93.

Bonaccorso, F. S., A. Arends, M. Genoud, D. Cantoni, and T. Morton. 1992. Thermal ecology of moustached and ghost-faced bats (Mormoopidae) in Venezuela. *Journal of Mammalogy* 73:365–78.

Bond, R. M., and G. A. Seaman. 1958. Notes on a colony of *Brachyphylla cavernarum. Journal of Mammalogy* 39:150–51.

Brass, D. A. 1994. *Rabies in Bats: Natural History and Public Health Implications.* Ridgefield, Connecticut: Livia Press.

Breckon, G. J. 2000. Revision of the flora of Desecheo Island, Puerto Rico. *Caribbean Journal of Science* 36:177–209.

Brigham, R. M., and M. B. Fenton. 1986. The influence of roost closure on the roosting and foraging behaviour of *Eptesicus fuscus* (Chiroptera: Vespertilionidae). *Canadian Journal of Zoology* 64:1128–33.

Brokaw, N. V. L., and J. S. Grear. 1991. Forest structure before and after Hurricane Hugo at three elevations in the Luquillo Mountains, Puerto Rico. *Biotropica* 23:386–92.

Brooke, A. P. 1994. Diet of the fishing bat, *Noctilio leporinus* (Chiroptera: Noctilionidae). *Journal of Mammalogy* 75:212–18.

————. 1997. Social organization and foraging behaviour of the fishing bat, *Noctilio leporinus* (Chiroptera: Noctilionidae). *Ethology* 103:421–36.

Brooke, A. P., and D. M. Decker. 1996. Lipid compounds in secretions of the fishing bat, *Noctilio leporinus* (Chiroptera: Noctilionidae). *Journal of Chemical Ecology* 22:1411–28.

Buden, D. W. 1976. A review of the bats of the West Indian genus *Erophylla. Proceedings of the Biological Society of Washington* 89:1–16.

————. 1977. First records of bats of the genus *Brachyphylla* from the Caicos Islands, with notes on geographic variation. *Journal of Mammalogy* 58:221–25.

Burnett, C. D., and T. H. Kunz. 1982. Growth rates and age estimation in *Eptesicus fuscus* and comparisons with *Myotis lucifugus. Journal of Mammalogy* 63:33–41.

Carvajal Zamora, J. R. 1977. Isolation of *Histoplasma capsulatum* from tissues of bats captured in the Aguas Buenas Caves, Aguas Buenas, Puerto Rico. *Mycopathologia* 60:167–69.

Chase, J., M. Y. Small, E. A. Weiss, D. Sharma, and S. Sharma. 1991. Crepuscular activity of *Molossus molossus*. *Journal of Mammalogy* 72:414–18.

Choate, J. R., and E. C. Birney. 1968. Sub-Recent Insectivora and Chiroptera from Puerto Rico, with description of a new bat of the genus *Stenoderma*. *Journal of Mammalogy* 49:400–12.

Clark, D. R. 1981. *Bats and Environmental Contaminants: A Review.* Special Scientific Report—Wildlife, 235. U.S. Department of the Interior, Fish and Wildlife Service. 27 pp.

Clark, D. R., Jr., and R. F. Shore. 2001. Chiroptera. Pp. 159–214 in *Ecotoxicology of Wild Mammals,* ed. R. F. Shore and B. A. Rattner. London: John Wiley and Sons.

Colón, J. A. 1977. Climatología. Pp. 47–122 in *Geovisión de Puerto Rico,* ed. M. T. B Galiñanes. Río Piedras: Editorial Universitaria, Universidad de Puerto Rico.

Conde Costas, C., and C. Gonzalez. 1990. Las cuevas y cavernas en el bosque xerofítico de Guánica. *Acta Científica* 4:113–26.

Conn, D. B., and S. A. Marshall. 1991. Microdistributions of scavenging flies in relation to detritus and guano deposits in a Kentucky bat cave. *Entomological News* 102:127–29.

Constantine, D. G. 1988. Health precautions for bat researchers. Pp. 491–528 in *Ecological and Behavioral Methods for the Study of Bats,* ed. T. H. Kunz. Washington, D.C.: Smithsonian Institution Press.

Culver, D. C. 1982. *Cave Life: Evolution and Ecology.* Cambridge, Mass.: Harvard University Press.

Daniel, M. J. 1979. The New Zealand short-tailed bat, *Mystacina tuberculata*: A review of present knowledge. *New Zealand Journal of Zoology* 6:357–70.

Dávalos, L. M., and R. Erikson. 2003. New and noteworthy records from ten Jamaican bat caves. *Caribbean Journal of Science* 39:140–44.

Davis, R. B., C. F. Herreid II, and H. L. Short. 1962. Mexican free-tailed bats in Texas. *Ecological Monographs* 32:311–46.

Desmarest, A. G. 1820. Mammalogie, ou description des espèces de mammifères, Paris, 1:1–276 (not seen; cited in Genoways and Baker, 1972).

Devoe, N. N. 1990. Differential seeding and regeneration rates in openings and under closed canopy in subtropical wet forests. Ph.D. diss., Yale University, New Haven, Connecticut.

Díaz-Díaz, C. 1983. *Compendio enciclopedico de los Recursos Naturales,* vol. 1: *Los mamíferos de Puerto Rico.* San Juan: Departamento de Recursos Naturales, Area de Investigaciones Científicas. 221 pp.

Dobson, G. E. 1878. *Catalogue of the Chiroptera in the Collection of the British Museum.* London: British Museum.

Dolan, P. G. 1989. Systematics of Middle American mastiff bats of the genus *Molossus.* Special Publication 29. The Museum, Texas Tech University. 71 pp.

Ewel, J. J., and J. L. Whitmore. 1973. *The Ecological Life Zones of Puerto Rico and the U.S. Virgin Islands.* Research Paper ITF-18. U.S. Department of Agriculture, Forest Service, Institute of Tropical Forestry. 72 pp.

Faaborg, J. 1977. Metabolic rates, resources, and the occurrence of nonpasserines in terrestrial avian communities. *American Naturalist* 111:903–16.

Fabian, M. E., and R. Vera-Marques. 1989. Contribuçao ao conhecimento da biología reprodutiva de *Molossus molossus* (Pallas, 1766) (Chiroptera: Molossidae). *Revista Brasileira de Zoologia* 6:603–10.

Fain, A., G. Anastos, J. Camin, and D. Johnston. 1967. Notes on the genus *Spelaeorhynchus:* Description of *S. praecursor* Neumann and of two new species. *Acarologia* 9:535–56.

Feldhamer, G. A., L. C. Drickamer, S. H. Vessey, and J. F. Merritt. 2004. *Mammalogy: Adaptation, Diversity, and Ecology.* New York: McGraw-Hill.

Fenton, M. B. 1982. Echolocation, insect hearing, and feeding ecology of insectivorous bats. Pp. 261–86 in *Ecology of Bats,* ed. T. H. Kunz. New York: Plenum Press.

Fenton, M. B., and G. P. Bell. 1981. Recognition of species of insectivorous bats by their echolocation calls. *Journal of Mammalogy* 62:233–43.

Fenton, M. B., and T. H. Kunz. 1977. Movements and behavior. Pp. 351–64 in *Biology of Bats of the New World Family Phyllostomatidae, Part II,* ed. R. J. Baker, J. K. Jones, Jr., and D. C. Carter. Special Publication 13. The Museum, Texas Tech University. 364 pp.

Fernández, D. S., and N. Fetcher. 1991. Changes in light availability following Hurricane Hugo in a subtropical montane forest in Puerto Rico. *Biotropica* 23:393–99.

Fernández Méndez, E. 1981. *Crónicas de Puerto Rico: Desde la conquista hasta nuestros dias (1493–1955).* Río Piedras: Editorial Universitaria, Universidad de Puerto Rico.

Fleming, T. H. 1971. *Artibeus jamaicensis:* Delayed embryonic development in a neotropical bat. *Science* 171:402–404.

———. 1982. Foraging strategies of plant-visiting bats. Pp. 287-325 in *Ecology of Bats,* ed. T. H. Kunz. New York: Plenum Press.

————. 1988. *The Short-Tailed Fruit Bat: A Study in Plant–Animal Interactions.* Chicago: University of Chicago Press.

Frank, E. F. 1998. History of the guano mining industry, Isla de Mona, Puerto Rico. *Journal of Cave and Karst Studies* 60:121–25.

French, B. 2002. Captive longevity record for *Eptesicus fuscus. Bat Research News* 42:198.

Gannon, M. R. 1991. Foraging ecology, reproductive biology, and systematics of the red fig-eating bat (*Stenoderma rufum*) in the tabonuco rain forest of Puerto Rico. Ph.D. diss., Texas Tech University, Lubbock.

————. 2002. Survey of bats on Puerto Rico after Hurricane Georges, with special emphasis on *Stenoderma rufum.* Unpublished report. U.S. Department of Agriculture, Forest Service, Caribbean Field Office, Boquerón, Puerto Rico.

Gannon, M. R., and M. R. Willig. 1992. Bat reproduction in the Luquillo Experimental Forest of Puerto Rico. *Southwestern Naturalist* 37:414–19.

————. 1994a. The effects of Hurricane Hugo on the bats of the Luquillo Experimental Forest of Puerto Rico. *Biotropica* 26:320–31.

————. 1994b. Records of bat ectoparasites from the Luquillo Experimental Forest of Puerto Rico. *Caribbean Journal of Science* 30:281–83.

————. 1995. Ecology of ectoparasites from tropical bats. *Journal of Environmental Entomology* 24:1495–1503.

————. 1997. The effect of lunar illumination on movement and activity of the red fig-eating bat (*Stenoderma rufum*). *Biotropica* 29:525–29.

Gannon, M. R., M. R. Willig, and J. K. Jones, Jr. 1992. Morphometric variation, measurement error, and fluctuating asymmetry in the red fig-eating bat (*Stenoderma rufum*). *Texas Journal of Science* 44:389–404.

Garcia, A. G., C. E. Diez, and A. O. Alvarez. 2001. The impact of feral cats on Mona Island wildlife and recommendations for their control. *Caribbean Journal of Science* 37:107–108.

Gardner, A. L. 1977. Feeding habits. Pp. 293–350 in *Biology of Bats of the New World Family Phyllostomatidae, Part II,* ed. R. J. Baker, J. K. Jones, Jr., and D. C. Carter. Special Publication 13. The Museum, Texas Tech University. 364 pp.

Genoways, H. H., and R. J. Baker. 1972. *Stenoderma rufum. Mammalian Species* 18:1–4.

Geoffroy Saint-Hilaire, É. 1818. Description des mammifères qui se trouvent en Égypte. Pp. 99-166 in *Description de l'Égypte ou recueil des observations et des recherches qui ont été faites en Égypte pendant l'expédition de l'armér Française [1798–1801],* vol. 2: *Histoire Naturelle.* Paris. 218 pp.

————. 1824. Memoire sur une chauve-souris Americaine, formant une nouvelle espèce dans le genre *Nyctinome*. *Annales des Sciences Naturelles* (Paris) 1:337–47.

Gerell, R., and K. Lundberg. 1985. Social organization in the bat *Pipistrellus pipistrellus*. *Behavioral Ecology and Sociobiology* 16:177–84.

Goodwin, G. G. 1970. The ecology of Jamaican bats. *Journal of Mammalogy* 51:571–79.

Gray, J. E. 1834. Characters of a new genus of bats (*Brachyphylla*), obtained by the society from the collection of the late Rev. Lansdown Guilding. *Proceedings of the Zoological Society of London* 1834:122–23.

———. 1843. (Letter addressed to the curator). *Proceedings of the Zoological Society of London* 1843:50.

Greenhall, A. M. 1982. *House Bat Management*. Resource Publication 143. U.S. Fish and Wildlife Service. 33 pp.

Greenhall, A. M., and U. Schmidt (eds.). 1988. *Natural History of Vampire Bats*. Boca Raton, Florida.: CRC Press.

Griffin, D. R., J. H. Friend, and F. A. Webster. 1960. The echolocation of flying insects by bats. *Animal Behaviour* 8:141–54.

Griffiths, T. A., and D. Klingener. 1988. On the distribution of Greater Antillean bats. *Biotropica* 20:240–51.

Gundlach, J. 1840. Beschreibung von vier auf Cuba gefangenen Fledermäusen. *Archives für Naturgeschichte* 6:356–58 (not seen; cited in Silva-Taboada, 1976).

Gundlach, J. C. 1861. Eine neue von Hrn. Dr. Gundlach beschriebene Gattung von Flederthieren aus Cuba. *Monatsbericht der Königlich Preussischen Akademie der Wissenschaften zu Berlin*, 1861:817–19 (not seen, cited in Baker et al., 1978).

Gutierrez, M., and A. Aoki. 1973. Fine structure of the gular gland of the free-tailed bat *Tadarida brasiliensis*. *Journal of Morphology* 141:293–306.

Hall, E. R. 1981. *The Mammals of North America*. 2nd ed. New York: John Wiley and Sons.

Hall, E. R., and J. W. Bee. 1960. The red fig-eating bat *Stenoderma rufum* found alive in the West Indies. *Mammalia* 24:67–75.

Hall, E. R., and J. R. Tamsitt, 1968. A new subspecies of the red fig-eating bat from Puerto Rico. Life Sciences, Occasional Paper 11. Royal Ontario Museum. 5 pp.

Hamilton, R. B., and D. T. Stalling. 1972. *Lasiurus borealis* with five young. *Journal of Mammalogy* 53:190.

Handley, C. O., Jr., D. E. Wilson, and A. L. Gardner. 1991. *Demography and*

Natural History of the Common Fruit Bat, Artibeus jamaicensis, *on Barro Colorado Island, Panamá.* Smithsonian Contributions to Zoology 511. 173 pp.

Hartley, D. J., and R. A. Suthers. 1987. The sound emission pattern and the acoustical role of the nose-leaf in the echolocating bat, *Carollia perspicillata. Journal of the Acoustic Society of America* 82:1892–1900.

Hawkins, C. C. 1998. Impact of a subsidized exotic predator on native biota: Effect of house cats (*Felis catus*) on California birds and rodents. Ph.D. diss., Texas A&M University, College Station.

Hayssen, V., and T. H. Kunz. 1996. Allometry of litter mass in bats: Maternal size, wing morphology, and phylogeny. *Journal of Mammalogy* 77:476–90.

Heath, P., and J. B. Schneewind (eds.). 1997. *Lectures on Ethics: Immanuel Kant.* New York: Cambridge University Press.

Heatwole, H., L. Kelts, R. Levins, and F. Torres. 1963. Faunal notes on Culebra Island, Puerto Rico. *Caribbean Journal of Science* 3:29–30.

Heatwole, H., J. F. Arroyo-Salaman, and G. Hernandez. 1964. Albinism in the bat *Molossus fortis. Journal of Mammalogy* 45:476.

Heithaus, E. R. 1982. Coevolution between bats and plants. Pp. 327–67 in *Ecology of Bats,* ed. T. H. Kunz. New York: Plenum Press.

Helmer, E. H., O. Ramos, T del M. López, M. Quiñones, and W. Diaz. 2002. Mapping the forest type and land cover of Puerto Rico, a component of the Caribbean biodiversity hotspot. *Caribbean Journal of Science* 38:165–38.

Herd, R. M. 1983. *Pteronotus parnellii. Mammalian Species* 209:1–5.

Hickey, M. B. C., and M. B. Fenton. 1990. Foraging by red bats (*Lasiurus borealis*): Do intraspecific chases mean territoriality? *Canadian Journal of Zoology* 68:2477–82.

Holdridge, L. R. 1947. Determination of world plant formations from simple climatic data. *Science* 105:367–68.

———. 1967. *Life Zone Ecology.* Rev. ed. San Jose, Costa Rica: Tropical Science Center.

Homan, J. A., and J. K. Jones, Jr. 1975. *Monophyllus redmani. Mammalian Species* 57:1–2.

Hood, C. S., and J. K. Jones, Jr. 1984. *Noctilio leporinus. Mammalian Species* 216:1–7.

Hoofer, S. R., and R. A. Van Den Bussche. 2003. Molecular phylogenetics of the chiropteran family Vespertilionidae. *Acta Chiropterologica* 5 (supplement): 1–63.

Horst, G. R., D. B. Hoagland, and C. W. Kilpatrick. 2001. The mongoose in the West Indies: The biogeography and population biology of an introduced species. Pp. 409–24 in *Biogeography of the West Indies: Patterns and Perspectives,* ed. C. A. Woods and F. E. Sergile. Boca Raton, Florida: CRC Press.

Howell, D. J., and J. Pylka. 1977. Why do bats hang upside down? A biomechanical hypothesis. *Journal of Theoretical Biology* 69:625–31.

Humphrey, S. R. 1975. Nursery roosts and community diversity of nearctic bats. *Journal of Mammalogy* 56:321–46.

Husson, A. M. 1962. The bats of Suriname. *Zoologische Vehandelingen* 58:1–282.

Hutchinson, G. E. 1950. *Survey of Existing Knowledge of Biogeochemistry,* vol. 3: *The Biogeochemistry of Vertebrate Excretion.* Bulletin 96. American Museum of Natural History. 554 pp.

Hutchinson, J. T., and M. J. Lacki. 2000. Selection of day roosts by red bats in mixed mesophytic forests. *Journal of Wildlife Management* 64:87–94.

Hutson, A. M., S. P. Mickleburgh, and P. A. Racey. 2001. *Microchiropteran Bats: Global Status Survey and Conservation Action Plan.* Gland, Switzerland: IUCN.

Jackson, H. H. T. 1916. A new bat from Porto Rico. *Proceedings of the Biological Society of Washington* 29:37–38.

Jackson, W. B. 1982. Norway rat and allies. Pp. 1077–88 in *Wild Mammals of North America,* ed. J. A. Chapman and G. A. Feldhamer. Baltimore, Maryland: Johns Hopkins University Press.

Jaeger, E. C. 1955. *A Source-Book of Biological Names and Terms.* 3rd ed. Springfield, Illinois: Charles C. Thomas.

Janzen, D. H. 1973. Sweep samples of tropical foliage insects: effects of seasons, vegetation types, elevation, time of day, and insularity. *Ecology* 54:687–702.

Jiménez de Wagenheim, O. 1998. *Puerto Rico: An Interpretive History from Pre-Columbian Times to 1900.* Princeton, New Jersey: Markus Weiner Publishers.

Jones, G., and J. Rydell. 1994. Foraging strategy and predation risk as factors influencing emergence time in echolocating bats. *Philosophical Transactions of the Royal Society of London,* B 346:445–55.

———. 2003. Attack and defense: Interactions between echolocating bats and their insect prey. Pp. 301–45 in *Bat Ecology,* ed. T. H. Kunz and M. B. Fenton. Chicago: University of Chicago Press.

Jones, J. K., Jr., H. H. Genoways, and R. J. Baker. 1971. Morphological variation in *Stenoderma rufum. Journal of Mammalogy* 52:244–47.

Jones, K. E., and A. MacLarnon. 2001. Bat life histories: Testing models of mammalian life-history evolution. *Evolutionary Ecology Research* 3:465–76.

Jones, K. E., and A. Purvis. 1997. An optimum body size for mammals? Comparative evidence from bats. *Functional Ecology* 11:751–56.

Jones, K. E., K. E. Barlow, N. Vaughan, A. Rodríguez-Durán, and M. R. Gannon.

2001. Short-term impacts of extreme environmental disturbances on the bats of Puerto Rico. *Animal Conservation* 4:59–66.

Jones, K. E., A. Purvis, A. MacLarnon, O. R. P. Bininda-Emonds, and N. B. Simmons. 2002. A phylogenetic supertree of the bats (Mammalia: Chiroptera). *Biological Reviews* 77:223–59.

Kalko, E. K. V., E. A. Herre, and C. O. Handley. 1996. The relation of fig fruit syndromes to fruit eating bats in the New and Old World tropics. *Journal of Biogeography* 23:565–76.

Kalko, E. K., H.-U. Schnitzler, I. Kaipf, and D. Grinnell. 1998. Echolocation and foraging of the lesser bulldog bat, *Noctilio albiventris*: Preadaptations for piscivory? *Behavioral Ecology and Sociobiology* 42:305–19.

Kaufman, D. M., and M. R. Willig. 1998. Latitudinal patterns of mammalian species richness in the New World: Effects of sampling method and faunal group. *Journal of Biogeography* 25:795–805.

Keeley, B. W., and M. D. Tuttle. 1999. *Bats in American Bridges*. Resource Publication 4. Austin, Texas: Bat Conservation International. 41 pp.

Koenig, N. 1953. *A Comprehensive Agricultural Program for Puerto Rico.* Washington, D.C.: U.S. Department of Agriculture and Commonwealth of Puerto Rico.

Koopman, K. F. 1975. Bats of the Virgin Islands in relation to those of the Greater and Lesser Antilles. *American Museum Novitates* 2581:1–7.

———. 1989. A review and analysis of the bats of the West Indies. Pp. 635–44 *in Biogeography of the West Indies: Past, Present, and Future,* ed. C. A. Woods. Gainesville, Florida: Sandhill Crane Press.

———. 1993. Order Chiroptera. Pp. 137–241 in *Mammal Species of the World: A Taxonomic and Geographic Reference.* 2nd ed., ed. D. E. Wilson and D. M. Reeder. Washington, D.C.: Smithsonian Institution Press.

Kössl, M., E. Mora, F. Coro, and M. Vater. 1999. Two-toned echolocation calls from *Molossus molossus* in Cuba. *Journal of Mammalogy* 80:929–32.

Krebs, J. W., A. M. Mondul, C. E. Rupprecht, and J. E. Childs. 2001. Rabies surveillance in the United States during 2000. *Journal of the American Veterinary Medicine Association* 219:1687–99.

Krutzsch, P. H., and E. G. Crichton. 1985. Observations on the reproductive cycle of female *Molossus fortis* (Chiroptera: Molossidae) in Puerto Rico. *Journal of Zoology* (London) 207:137–50.

———. 1990. Observations on the reproductive anatomy of the male *Molossus fortis* (Chiroptera: Molossidae) with comments on the chronology of reproductive events. *Mammalia* 54:287–96.

Kunz, T. H. 1982. Roosting ecology of bats. Pp. 1–55 in *Ecology of Bats,* ed. T. H. Kunz. New York: Plenum Press.

Kunz, T. H., and C. A. Diaz. 1995. Folivory in fruit-eating bats, with new evidence from *Artibeus jamaicensis* (Chiroptera: Phyllostomidae). *Biotropica* 27:106–20.

Kunz, T. H., and D. P. Jones. 2000. *Pteropus vampyrus*. *Mammalian Species* 642:1–6.

Kunz, T. H., and A. Kurta. 1988. Capture methods and holding devices. Pp. 1–29 in *Ecological and Behavioral Methods for the Study of Bats,* ed. T. H. Kunz. Washington, D.C.: Smithsonian Institution Press.

Kunz, T. H., and L. F. Lumsden. 2003. Ecology of cavity and foliage roosting bats. Pp. 3–89 in *Bat Ecology,* ed. T. H. Kunz and M. B. Fenton. Chicago: University of Chicago Press.

Kunz, T. H., and S. K. Robson. 1995. Postnatal growth and development in the Mexican free-tailed bat (*Tadarida brasiliensis mexicana*): Birth size, growth rates, and age estimation. *Journal of Mammalogy* 76:769–83.

Kunz, T. H., P. V. August, and C. D. Burnett. 1983. Harem social organization in cave roosting *Artibeus jamaicensis* (Chiroptera: Phyllostomidae). *Biotropica* 15:133–38.

Kunz, T. H., M. S. Fujita, A. P. Brooke, and G. F. McCracken. 1994. Convergence in tent architecture and tent-making behavior among neotropical and paleotropical bats. *Journal of Mammalian Evolution* 2:57–78.

Kunz, T. H., O. T. Oftedal, S. K. Robson, M. B. Kretzmann, and C. Kirk. 1995. Changes in milk composition during lactation in three species of insectivorous bats. *Journal of Comparative Physiology,* B 164:543–51.

Kurta, A. 1982. Flight patterns of *Eptesicus fuscus* and *Myotis lucifugus* over a stream. *Journal of Mammalogy* 63:335–37.

———. 1985. External insulation available to a non-nesting mammal, the little brown bat (*Myotis lucifugus*). *Comparative Biochemistry and Physiology* 82A:413–20.

Kurta, A., and R. H. Baker. 1990. *Eptesicus fuscus*. *Mammalian Species* 356:1–10.

Kurta, A., and M. Ferkin. 1991. The correlation between demography and metabolic rate: A test using the beach vole (*Microtus breweri*) and the meadow vole (*Microtus pennsylvanicus*). *Oecologia* 87:102–105.

Kurta, A., and T. H. Kunz. 1987. Size of bats at birth and maternal investment during pregnancy. *Symposia of the Zoological Society of London* 57:79–107.

Kurta, A., and M. E. Stewart. 1991. Parturition in the silver-haired bat, *Lasionycteris noctivagans,* with a description of the neonates. *Canadian Field-Naturalist* 104:598–600.

Kurta, A., T. H. Kunz, and K. A. Nagy. 1990. Energetics and water flux of free-ranging big brown bats (*Eptesicus fuscus*) during pregnancy and lactation. *Journal of Mammalogy* 71:59–65.

Kurta, A., G. P. Bell, K. A. Nagy, and T. H. Kunz. 1989. Energetics of pregnancy and lactation in free-ranging little brown bats (*Myotis lucifugus*). *Physiological Zoology* 62:804–18.

Lancaster, W. C., and E. K. V. Kalko. 1996. *Mormoops blainvillii. Mammalian Species* 544:1–5.

Lancaster, W. C., O. W. Henson, Jr., and A. W. Keating. 1995. Respiratory muscle activity in relation to vocalization in flying bats. *Journal of Experimental Biology* 198:175–91.

Larue, D. K. 1994. Puerto Rico and the Virgin Islands. Pp. 151–65 in *Caribbean Geology: An Introduction.* Kingston, Jamaica: University of the West Indies Publishers' Association.

Lawrence, B. 1977. Dogs from the Dominican Republic. *Cuadernos del Cendia* 8:3–19.

Leach, W. E. 1821. The characters of seven genera of bats with foliaceous appendages to the nose. *Transactions of the Linnean Society of London* 12:73–82.

Lewis-Oritt, N., R. A. Van Den Bussche, and R. J. Baker. 2001. Molecular evidence for evolution of piscivory in *Noctilio* (Chiroptera: Noctilionidae). *Journal of Mammalogy* 82:748–59.

Linnaeus, C. 1758. *Systema Naturae per Regna Tria Naturae, Secundum Classis, Ordines, Genera, Species cum Characteribus, Differentiis Synonymis, Locis.* 10th ed. 1:1–824 (not seen; cited in Koopman, 1993).

Liogier, H. A., and L. F. Martorell. 1982. *Flora of Puerto Rico and Adjacent Islands: A Systematic Synopsis.* Río Piedras: Editorial de la Universidad de Puerto Rico.

Little, E. L., and F. H. Wadsworth. 1964. *Common Trees of Puerto Rico and the Virgin Islands.* Handbook 249. U.S. Department of Agriculture, Forest Service. 548 pp.

López-González, C., and S. J. Presley. 2001. Taxonomic status of *Molossus bondae* J. A. Allen, 1904 (Chiroptera: Molossidae), with description of a new subspecies. *Journal of Mammalogy* 82:760–74.

Lugo, A. E., and E. Helmer. 2004. Emerging forests on abandoned land: Puerto Rico's new forests. *Forest and Ecology Management* 190:154–61.

Lugo, A. E., L. Miranda Castro, A. Vale, T del Mar López, E. Hernández Prieto, A. Garcia Martino, A. R. Puente Rolón, A. G. Tossas, D. A. MacFaelane, T. Miller, A. Rodríguez, J. Lundberg, J. Thomlinson, J. Colón, J. H. Schellekens, O. Ramos, and E. Helmer. 2001. *Puerto Rican Karst: A Vital Resource.* Technical Report WO-65 U.S. Department of Agriculture, Forest Service. 100 pp.

Ma, J., G. Jones, S. Zhang, J. Shen, W. Metzner, L. Zhang, and B. Liang. 2003. Dietary analysis confirms that Rickett's big-footed bat (*Myotis ricketti*) is a piscivore. *Journal of Zoology* (London) 261:245–48.

MacArthur, R. H., and E. O. Wilson. 1967. *The Theory of Island Biogeography.* Monographs in population biology 1. Princeton, New Jersey: Princeton University Press.

Macías, S., and E. C. Mora. 2003. Variation of echolocation calls of *Pteronotus quadridens* (Chiroptera: Mormoopidae) in Cuba. *Journal of Mammalogy* 84:1428–36.

Mager, K. J., and T. A. Nelson. 2001. Roost site selection by eastern red bats (*Lasiurus borealis*). *American Midland Naturalist* 145:120–26.

Marinho-Filho, J., and I. Sazima. 1998. Brazilian bats and conservation biology: A first survey. Pp. 282–94 in *Bat Biology and Conservation,* ed. T. H. Kunz and P. A. Racey. Washington, D.C.: Smithsonian Institution Press.

Mattson, P., G. Draper, and J. F. Lewis. 1990. Puerto Rico and the Virgin Islands. Pp. 112–20 in *The Geology of North America,* vol. H: *The Caribbean Region,* ed. G. Dengo and J. E. Case. New York: Geological Society of America.

McCracken, G. F. 1996. Bats aloft: A study of high altitude feeding. *Bats* 14:7–10.

———. 1999. Brazilian free-tailed bat *Tadarida brasilensis.* Pp. 127–29 in *The Smithsonian Book of North American Mammals,* ed. S. Ruff and D. E. Wilson. Washington, D.C.: Smithsonian Institution Press.

McCracken, G. F., and M. K. Gustin. 1991. Nursing behavior in Mexican free-tailed bat maternity colonies. *Ethology* 89:305–21.

McCracken, G. F., and G. S. Wilkinson. 2000. Bat mating systems. Pp. 321–62 in *Reproductive Biology of Bats,* ed. E. G. Crichton and P. H. Krutzsch. New York: Academic Press.

McNab, B. K. 1982. Evolutionary alternatives in the physiological ecology of bats. Pp. 151–200 in *Ecology of Bats,* ed. T. H. Kunz. New York: Plenum Press.

———. 1994. Resource use and the survival of land and freshwater vertebrates on oceanic islands. *American Naturalist* 144:643–60.

———. 2001. Functional adaptations to island life in the West Indies. Pp. 55–62 in *Biogeography of the West Indies: Patterns and Perspectives,* ed. C. A. Woods and F. E. Sergile. Boca Raton, Florida: CRC Press.

Menzel, M. A., T. C. Carter, B. R. Chapman, and J. Laerm. 1998. Quantitative comparison of tree roosts used by red bats (*Lasiurus borealis*) and Seminole bats (*L. seminolus*). *Canadian Journal of Zoology* 76:630–34.

Messenger, S. L., C. E. Rupprecht, and J. S. Smith. 2003. Bats, emerging virus infections, and the rabies paradigm. Pp. 622–79 in *Bat Ecology,* ed. T. H. Kunz and M. B. Fenton. Chicago: University of Chicago Press.

Miller, G. S., Jr. 1899. Two new glossophagine bats from the West Indies. *Proceedings of the Biological Society of Washington* 13:33–37.

———. 1900. The bats of the genus *Monophyllus*. *Proceedings of the Washington Academy of Science* 2:31–38.

———. 1913. Notes on the bats of the genus *Molossus*. *Proceedings of the U.S. National Museum* 46:85–92.

———. 1918. Mammals and reptiles collected by Theodoor de Boy in the Virgin Islands. *Proceedings of the United States National Museum* 54:507–11.

———. 1929. *Mammals Eaten by Indians, Owls, and Spaniards in the Coast Region of the Dominican Republic.* Smithsonian Miscellaneous Collections 82. 16 pp.

———. 1931. The red bats of the Greater Antilles. *Journal of Mammalogy* 12:409–10.

Moorman, C. E., K. R. Russell, M. A. Menzel, S. M. Lohr, J. E. Ellenberger, and D. H. Van Lear. 1999. Bats roosting in deciduous leaf litter. *Bat Research News* 40:74–75.

Morales, J. C., and J. W. Bickham. 1995. Molecular systematics of the genus *Lasiurus* (Chiroptera: Vespertilionidae) based on restriction-site maps of the mitichondrial ribosomal genes. *Journal of Mammalogy* 76:730–49.

Morgan, G. S. 2001. Patterns of extinction in West Indian bats. Pp. 369–408 in *Biogeography of the West Indies: Patterns and Perspectives*, ed. C. A. Woods and F. E. Sergile. Boca Raton, Florida: CRC Press.

Morgan, G. S., and C. A. Woods. 1986. Extinction and zoogeography of West Indian land mammals. *Biological Journal of the Linnean Society* 28:167–203.

Morrison, D. W. 1978a. Influence of habitat on the foraging distances of the fruit bat *Artibeus jamaicensis*. *Journal of Mammalogy* 59:622–24.

———. 1978b. Foraging ecology and energetics of the frugivorous bat *Artibeus jamaicensis*. *Ecology* 59:716–23.

———. 1978c. Lunar phobia in a neotropical fruit bat, *Artibeus jamaicensis* (Chiroptera: Phyllostomidae). *Animal Behaviour* 26:852–55.

Müller, P. L. S. 1776. *Mit einer ausfürlichen erklärung ausgefertiget. Des ritters Carl von Linne . . . Vollständigen natursystems supplements and register-ban über aller sechs theile oder classen des thierreichs.* Nürnberg, Germany: G. N. Raspe (not seen; cited in Shump and Shump, 1972).

Murphy, M. 1991. Fables, fear, and creatures that fly in the night. *Bats* 9:15.

Murray, P. F., and T. Strickler. 1975. Notes on the structure and function of cheek pouches within the Chiroptera. *Journal of Mammalogy* 56:673–76.

Myers, P. 1977. *Patterns of Reproduction of Four Species of Vespertilionid Bats in Paraguay.* University of California Publications in Zoology 107. 41 pp.

Nagorsen, D. W., and R. L. Peterson. 1975. Karyotypes of six species of bats

(Chiroptera) from the Dominican Republic. *Life Sciences Occasional Papers* 28. Royal Ontario Museum. 8 pp.

Nellis, D. W. 1971. Additions to the natural history of *Brachyphylla* (Chiroptera). *Caribbean Journal of Science* 11:91.

Nellis, D. W., and C. P. Ehle. 1977. Observations on the behavior of *Brachyphylla cavernarum* (Chiroptera) in the Virgin Islands. *Mammalia* 41:403–409.

Nieves-Rivera, Á. 2003. Mycological survey of Río Camuy Caves State Park, Puerto Rico. *Journal of Cave and Karst Studies* 65:22–28.

Nowak, R. M. 1999. *Walker's Mammals of the World,* vol. 1. 6th ed. Baltimore, Maryland: Johns Hopkins University Press.

O'Farrell, M. J., and B. W. Miller. 1999. Use of vocal signatures for the inventory of free-flying neotropical bats. *Biotropica* 31:507–16.

O'Farrell, M. J., B. W. Miller, and W. L. Gannon. 1999. Qualitative identification of free-flying bats using the Anabat detector. *Journal of Mammalogy* 80:11–23.

Olson, S. L. 1982. Biological archaeology in the West Indies. *Florida Anthropologist* 35:162–68.

Ortega, J., and I. Castro-Arellano. 2001. *Artibeus jamaicensis. Mammalian Species* 662:1–9.

Ortiz, P. R. 1989. A summary of conservation trends in Puerto Rico. Pp. 851–54 in *Biogeography of the West Indies: Past, Present, and Future,* ed. C. A. Woods. Gainesville, Florida: Sandhill Crane Press.

Owen, R. D. 1988. Phenetic analyses of the bat subfamily Stenodermatinae (Mammalia: Chiroptera). *Journal of Mammalogy* 69:795–810.

Palisot de Beauvois, A. M. F. J. 1796. *Catalog raisonne du muséum de Mr. C. W. Peale.* Philadelphia, Pennsylvania: Parent.

Pallas, P. S. 1766. *Miscellanea zoological quibus novae imprimis atque obscurae animalium species describuntur et observationibus iconibusque illustratum.* Hagae Company (not seen; cited in Husson, 1962).

Palmerim, J. M., and L. Rodrígues. 1992. *Plano nacional de conservação dos morcegos cavernícolas.* Estudos de Biologia e Conservação da Natureza, no. 8. Lisbon, Portugal.

Parsons, S., A. M. Boonman, and M. K. Obrist. 2000. Advantages and disadvantages of techniques for transforming and analyzing chiropteran echolocation calls. *Journal of Mammalogy* 81:927–38.

Patterson, B. D., M. R. Willig, and R. D. Stevens. 2003. Trophic strategies, niche partitioning, and patterns of ecological organization. Pp. 536–79 in *Bat Ecology,* ed. T. H. Kunz and M. B. Fenton. Chicago: University of Chicago Press.

Peck, S. B. 1974. The invertebrate fauna of tropical American caves, part 2: Puerto Rico, an ecological and zoogeographic analysis. *Biotropica* 6:14–31.

———. 1981. The subterranean fauna and conservation of Mona Island (Puerto Rico): A Caribbean karst environment. *National Speleological Society Bulletin* 43:59–68.

Pedersen, S. C. 2000. Sub-lethal pathology correlated with volcanic eruptions on Montserrat, BWI. *Bat Research News* 41:134.

———. 2002. Montserrat redux-recovery after a seven-year itch? *Bat Research News* 43:175.

Pedersen, S. C., H. H. Genoways, and P. W. Freeman. 1996. Notes on bats from Montserrat (Lesser Antilles) with comments concerning the effects of Hurricane Hugo. *Caribbean Journal of Science* 32:206–13.

Pedersen, S. C., H. H. Genoways, M. N. Norton, J. W. Johnson, and S. Courts. 2003. Bats of Nevis, northern Lesser Antilles. *Acta Chiropterologica* 5:252–67.

Petit, S. 1996. The status of bats on Curaçao. *Biological Conservation* 77:27–31.

Phillips, C. J. 1971. The dentition of glossophagine bats: Development, morphological characteristics, variation, pathology, and evolution. Miscellaneous Publication 54. Museum of Natural History, University of Kansas. 138 pp.

Picó, R. 1950. *The Geographic Regions of Puerto Rico.* Río Piedras: University of Puerto Rico Press.

———. 1974. *The Geography of Puerto Rico.* Chicago: Aldine Publishing Company.

Pine, R. H. 1980. Keys to the bats of Jamaica, Hispaniola and Puerto Rico based on gross external characters. *Caribbean Journal of Science* 15:9–11.

Pollak, G., and O. W. Henson, Jr. 1973. Specialized functional aspects of the middle ear muscles in the bat, *Chilonycteris parnellii. Journal of Comparative Physiology* 84:167–74.

Pollak, G., O. W. Henson, Jr., and A. Novick. 1972. Cochlear microphonic audiograms in the "pure tone" bat *Chilonycteris parnellii parnellii. Science* 176:66–68.

Pregill, G. K., and S. L. Olson. 1981. Zoogeography of West Indian vertebrates in relation to Pleistocene climatic cycles. *Annual Review of Ecology and Systematics* 12:75–98.

Price, J. L., and C. O. R. Everard. 1977. Rabies virus and antibody in bats in Grenada and Trinidad. *Journal of Wildlife Diseases* 13:131–34.

Racey, P. A., and A. C. Entwistle. 2003. Conservation ecology of bats. Pp. 680–743 in *Bat Ecology,* ed. T. H. Kunz and M. B. Fenton. Chicago: University of Chicago Press.

Raffaele, H. A., M. J. Velez, R. Cotte, J. J. Whelan, E. R. Keil, and W. Cupiano. 1973. *Rare and Endangered Animals of Puerto Rico and the Virgin Islands.* San Juan, Puerto Rico: Fondo Educativo Interamericano.

Rafinesque, C. S. 1820. *Annals of Nature or Annual Synopsis of New Genera and Species of Animals, Plants, &c. Discovered in North America.* Lexington, Kentucky: Thomas Smith.

Ray, C. E. 1964. The taxonomic status of *Heptaxodon* and dental ontogeny in *Elasmodontomys* and *Amblyrhiza*. *Bulletin of the Museum of Comparative Zoology* 131:107–27.

Reid, F. A. 1997. *A Field Guide to the Mammals of Central America and Southeast Mexico.* New York: Oxford University Press.

Rezsutek, M., and G. N. Cameron. 1993. *Mormoops megalophylla. Mammalian Species* 448:1–5.

Rivera-Marchand, B. 2001. Pollination biology of *Pilosocereus royennii* L. (Cactaceae) in Guánica State Forest. M.S. thesis, University of Puerto Rico, Río Piedras.

Rivera-Marchand, B., and A. Rodríguez-Durán. 2001. Preliminary observations on the renal adaptations of bats roosting in hot caves in Puerto Rico. *Caribbean Journal of Science* 37:272–74.

Rodriguez, G. A., and D. P. Reagan. 1984. Bat predation by the Puerto Rican boa, *Epicrates inornatus. Copeia* 1984:219–20.

Rodríguez-Durán, A. 1984. Community structure of a bat colony at Cueva Cucaracha. M.S. thesis, University of Puerto Rico, Mayagüez.

———. 1995. Metabolic rates and thermal conductance in four species of neotropical bats roosting in hot caves. *Comparative Biochemistry and Physiology* 110A:347–55.

———. 1996. Foraging ecology of the Puerto Rican boa (*Epicrates inornatus*): Bat predation, carrion feeding, and piracy. *Journal of Herpetology* 30:533–36.

———. 1998. Nonrandom aggregations and distribution of the cave bats of Puerto Rico. *Journal of Mammalogy* 79:141–46.

———. 1999. First record of a reproductive *Lasiurus borealis minor* (Miller) from Puerto Rico (Chiroptera). *Caribbean Journal of Science* 35:143–44.

———. 2002. Evaluation of the status of bat populations in western Vieques: Recommendations for a wildlife refuge management plan. Unpublished report. Puerto Rican Conservation Trust, San Juan.

———. 2002. Los murciélagos en las culturas pre-columbinas de Puerto Rico. *Focus* 1:15–18.

Rodríguez-Durán, A., and T. H. Kunz. 1992. *Pteronotus quadridens. Mammalian Species* 395:1–4.

————. 2001. Biogeography of West Indian bats: An ecological perspective. Pp. 355–68 in *Biogeography of the West Indies: Patterns and Perspectives,* ed. C. A. Woods and F. E. Sergile. Boca Raton, Florida: CRC Press.

Rodríguez-Durán, A., and A. R. Lewis. 1985. Seasonal predation by merlins on sooty mustached bats in western Puerto Rico. *Biotropica* 17:71–74.

————. 1987. Patterns of population size, diet, and activity time for a multispecies assemblage of bats at a cave in Puerto Rico. *Caribbean Journal of Science* 23:352–60.

Rodríguez-Durán, A., and J. A. Soto-Centeno. 2003. Temperature selection by tropical bats roosting in caves. *Journal of Thermal Biology* 28:465–68.

Rodríguez-Durán, A., and R. Vázquez. 2001. The bat *Artibeus jamaicensis* in Puerto Rico (West Indies): Seasonality of diet, activity, and effect of a hurricane. *Acta Chiropterologica* 3:53–61.

Rodríguez-Durán, A., A. R. Lewis, and Y. Montes. 1993. Skull morphology and diet of Antillean bat species. *Caribbean Journal of Science* 29:258–61.

Rudnick, A. 1960. A revision of the mites of the family Spinturnicidae (Acarina). *University of California Publications in Entomology* 17:157–283.

Rupprecht, C. E., C. A. Hanlon, and T. Hemachudha. 2002. Rabies re-examined. *Lancet, Infectious Diseases* 2:101–109.

Rutkowska, M. A. 1980. The helminthofauna of bats (Chiroptera) from Cuba. *Acta Parasitologica Polonica* 26:153–86.

Scatena, F. N., and M. C. Larsen. 1991. Physical aspects of Hurricane Hugo in Puerto Rico. *Biotropica* 23:317–23.

Schnitzler, H.-U., and K. V. Kalko. 1998. How echolocating bats approach and acquire food. Pp. 183–96 in *Bat Biology and Conservation,* ed. T. H. Kunz and P. A. Racey. Washington, D.C.: Smithsonian Institution Press.

Schnitzler, H.-U., E. Kalko, I. Kaipf, and A. D. Grinnell. 1994. Fishing and echolocation behavior of the greater bulldog bat, *Noctilio leporinus,* in the field. *Behavioral Ecology and Sociobiology* 35:327–45.

Schnitzler, H.-U., E. Kalko, I. Kaipf, and J. Mogdans. 1991. Comparative studies of echolocation and hunting behavior in the four species of mormoopid bats of Jamaica. *Bat Research News* 32:22–23.

Schutt, W. A., Jr. 1993. Digital morphology in the Chiroptera: The passive digital lock. *Acta Anatomica* 148:219–27.

Schutt, W. A., Jr., and N. A. Simmons. 2002. Morphological specializations in *Cheiromeles* (naked bulldog bats; Molossidae) and their possible role in quadrupedal locomotion. *Acta Chiropterologica* 3:225–35.

Schwartz, A., and J. K. Jones, Jr. 1967. Review of bats of the endemic Antillean genus *Monophyllus. Proceedings of the United States National Museum* 124:1–20.

Scogin, R. 1982. Dietary observations on the red fig-eating bat (*Stenoderma rufum*) in Puerto Rico. *Aliso* 10:259–61.

Secrest, M. F., M. R. Willig, and L. L. Peppers. 1996. The legacy of disturbance on habitat associations of terrestrial snails in the Luquillo Experimental Forest, Puerto Rico. *Biotropica* 28:502–14.

Sergile, F. E., and C. A. Woods. 2001. Status of conservation in Haiti: A 10-year retrospective. Pp. 547–60 in *Biogeography of the West Indies: Patterns and Perspectives,* ed. C. A. Woods and F. E. Sergile. Boca Raton, Florida: CRC Press.

Sherman, H. B. 1937. Breeding habits of the free-tailed bat. *Journal of Mammalogy* 18:176–87.

Shump, K. A., Jr., and A. U. Shump. 1982. *Lasiurus borealis. Mammalian Species* 183:1–6.

Siemers, B. M., C. Dietz, D. Nill, and H.-U. Schnitzler. 2001. *Myotis daubentonii* is able to catch small fish. *Acta Chiropterologica* 3:71–75.

Silva-Taboada, G. 1976. Historia y actualización taxonómica de algunas especies antillanas de murciélagos de los géneros *Pteronotus, Brachyphylla, Lasiurus,* y *Antrozous* (Mammalia: Chiroptera). *Poeyana* 153:1–24.

———. 1979. *Los murciélagos de Cuba.* Havana: Editorial de la Academia de Ciencias de Cuba.

Silva Taboada, G., and M. Herrada Libre. 1974. Primer caso comprobada de rabia en un murciélago cubano. *Poeyana* 126:1–5.

Silva-Taboada, G., and R. H. Pine. 1969. Morphological and behavioral evidence for the relationship between the bat genus *Brachyphylla* and the Phyllonycterinae. *Biotropica* 1:10–19.

Simmons, J. A., M. B. Fenton, and M. J. O'Farrell. 1979. Echolocation and pursuit of prey by bats. *Science* 203:16–21.

Simmons, N. B. 2005. Order Chiroptera. In *Mammal Species of the World: A Taxonomic and Geographic Reference.* 3rd ed., ed. D. E. Wilson and D. M Reeder. Baltimore, Maryland: Johns Hopkins University Press.

Simmons, N. B., and T. M. Conway. 2001. *Phylogenetic Relationships of Mormoopid Bats (Chiroptera: Mormoopidae) Based on Morphological Data.* Bulletin 258. American Museum of Natural History. 97 pp.

Smith, J. D. 1972. Systematics of the chiropteran family Mormoopidae. Miscellaneous Publication 56. Museum of Natural History, University of Kansas. 132 pp.

Snyder, N. F. R., J. W. Wiley, and C. B. Kepler. 1987. *The Parrots of Luquillo:*

Natural History and Conservation of the Puerto Rican Parrot. Los Angeles, California: Western Foundation of Vertebrate Zoology.

Soto-Centeno, J. A. 2004. Dietary ecology of two nectarivorous bats, *Erophylla sezekorni* and *Monophyllus redmani,* in Puerto Rico. M.S. thesis, Eastern Michigan University, Ypsilanti.

Soto-Centeno, J. A., and A. Kurta. 2003. Description of fetal and neonatal brown flower bats, *Erophylla sezekorni* (Chiroptera: Phyllostomidae). *Caribbean Journal of Science* 39:233–34.

Soto-Centeno, J. A., A. Rodríguez-Durán, and E. Cortes-Rosa. 2001. *Erophylla sezekorni* and *Brachyphylla cavernarum:* Diet of two phyllostomid bats in Puerto Rico. *Bat Research News* 42:180–81.

Speakman, J. R. 1995. Chiropteran nocturnality. *Symposia of the Zoological Society of London* 67:187–202.

Stadelmann, B., L. G. Herrera, J. Arroyo-Cabrales, J. J. Flores-Martinez, B. P. May, and M. Ruedi. 2004. Molecular systematics of the fishing bat *Myotis* (*Pizonyx*) *vivesi. Journal of Mammalogy* 85:133–139.

Starrett, A., and F. J. Rolle. 1962. A record of the genus *Lasiurus* from Puerto Rico. *Journal of Mammalogy* 44:264.

Studier, E. H., L. R. Beck, and R. G. Lindeborg. 1967. Tolerance and initial metabolic response to ammonia intoxication in selected bats and rodents. *Journal of Mammalogy* 48:564–72.

Studier, E. H., S. H. Sevick, D. M. Ridley, and D. E. Wilson. 1994. Mineral and nitrogen concentrations in feces of some neotropical bats. *Journal of Mammalogy* 75:674–80.

Suárez, W., and S. Díaz-Franco. 2003. A new fossil bat (Chiroptera: Phyllostomidae) from a Quaternary cave deposit in Cuba. *Caribbean Journal of Science* 39:371–77.

Swanepoel, P., and H. H. Genoways. 1978. Revision of the Antillean Bats of the Genus *Brachyphylla* (Mammalia: Phyllostomatidae). Bulletin 12. Carnegie Museum of Natural History. 53 pp.

———. 1983. *Brachyphylla cavernarum. Mammalian Species* 205:1–6.

Tamsitt, J. R. 1970. Comparative biochemistry and ecology of bats from the Puerto Rican region. *American Philosophical Society Yearbook* 1971:342–43.

Tamsitt, J. R., and I. Fox. 1970a. Records of bat ectoparasites from the Caribbean region (Siphonaptera, Acarina, Diptera). *Canadian Journal of Zoology* 48:1093–97.

———. 1970b. Mites of the family Listrophoridae in Puerto Rico. *Canadian Journal of Zoology* 48:398–99.

Tamsitt, J. R., and D. Valdivieso. 1966. Parturition in the red fig-eating bat, *Stenoderma rufum. Journal of Mammalogy* 47:352–53.

Tamsitt, J. R., and D. Valdivieso. 1970. Observations on bats and their ectoparasites. Pp. E123–E128 in *A Tropical Rain Forest,* ed. H. T. Odum and R. F. Pigeon. Oak Ridge, Tenn.: U.S. Atomic Energy Commission.

Teeling, E. C., O. Madsen, R. A. Van Den Bussche, W. W. de Jong, M. J. Stanhope, and M. S. Springer. 2002. Microbat paraphyly and the convergent evolution of a key innovation in Old World rhinolophid microbats. *Proceedings of the National Academy of Sciences* 99:1431–36.

Thies, W., E. K. V. Kalko, and H.-U. Schnitzler. 1998. The roles of echolocation and olfaction in two Neotropical fruit-eating bats, *Carollia perspicillata* and *C. castanea. Behavioral Ecology and Sociobiology* 42:397–409.

Thomas, K. R., and R. Thomas. 1974. Notes on *Stenoderma rufum* Desmarest. *Bat Research News* 15:24–25.

Thomlinson, J. R., M. I. Serrano, T. del M. Lopez, T. M. Aide, and J. Zimmerman. 1996. Land-use dynamics in a post-agricultural Puerto Rican landscape. *Biotropica* 28:525–36.

Thomson, S. C., A. P. Brooke, and J. R. Speakman. 1998. Diurnal activity in the Samoan flying fox, *Pteropus samoensis. Philosophical Transactions of the Royal Society of London,* B 353:1595–1606.

Timm, R. M., and H. H. Genoways. 2003. *West Indian Mammals from the Albert Schwartz Collection: Biological and Historical Information.* Scientific Paper 29. Natural History Museum, University of Kansas. 47 pp.

Trajano, E., and P. Gnaspini-Netto. 1994. Notes on the food webs in caves of southeastern Brazil. *Memoires de Biospeologie* 18:75–79.

Tuttle, M. D. 1994. The lives of Mexican free-tailed bats. *Bats* 12:6–14.

Tuttle, M. D., and D. L. Hensley. 1993. *The Bat House Builder's Handbook.* Austin, Texas: Bat Conservation International.

Tuttle, M. D., and S. J. Kern. 1981. *Bats and Public Health.* Contributions in Biology and Geology 48. Milwaukee Public Museum. 11 pp.

Tuttle, M. D., and D. Stevenson. 1982. Growth and survival of bats. Pp. 105–50 in *Ecology of Bats,* ed. T. H. Kunz. New York: Plenum Press.

Tuttle, M. D., and D. A. R. Taylor. 1998. *Bats and Mines.* Resource Publication 3. Austin, Texas: Bat Conservation International. 50 pp.

U.S. Department of Agriculture, Soil Conservation Service, and Commonwealth Department of Natural Resources. 1973. *Rare and Endangered Animal Species of Puerto Rico.* Committee report. San Juan, Puerto Rico.

———. 1975. *Rare and Endangered Plants of Puerto Rico.* Committee report. San Juan, Puerto Rico.

Van Den Bussche, R. A., S. R. Hoofer, and N. B. Simmons. 2002. Phylogenetic relationships of mormoopid bats using mitochondrial gene sequences and morphology. *Journal of Mammalogy* 83:40–48.

Van Den Bussche, R. A., R. J. Baker, H. A. Wichman, and M. J. Hamilton. 1993. Molecular phylogenetics of Stenodermatini bat genera: Congruence of data from nuclear and mitochondrial DNA. *Molecular Biology and Evolution* 10:944–59.

Varona, L. S. 1974. *Catálogo de los mamíferos vivientes y extinguidos de las Antillas.* Havana: Editorial de la Academia de Ciencias de Cuba.

Vaughan, N., and J. E. Hill. 1996. Bat (Chiroptera) diversity and abundance in banana plantations and rain forest, and three new records for St. Vincent, Lesser Antilles. *Mammalia* 60:441–47.

Vaughan, T. A. 1966. Morphology and flight characteristics of molossid bats. *Journal of Mammalogy* 47:249–60.

Vaughan, T. A. and G. C. Bateman. 1970. Functional morphology of the forelimb of mormoopid bats. *Journal of Mammalogy* 51:217–35.

Villa-C., B., and M. Canela-R. 1988. Man, gods, and legendary vampire bats. Pp. 234–40 in *Natural History of Vampire Bats,* ed. A. M. Greenhall and U. Schmidt. Boca Raton, Florida: CRC Press.

von Helversen, D., and O. von Helversen. 1999. Acoustic guide in bat-pollinated flower. *Nature* 398:759–60.

von Helversen, D., and Y. Winter. 2003. Glossophagine bats and their flowers: Costs and benefits for plants and pollinators. Pp. 346–97 in *Bat Ecology,* ed. T. H. Kunz and M. B. Fenton. Chicago: University of Chicago Press.

Vonhoff, M. J., and R. M. R. Barclay. 1997. Use of tree stumps as roosts by the western long-eared bat. *Journal of Wildlife Management* 61:674–84.

Wadsworth, F. H. 1949. The development of the forest land resources of the Luquillo Mountains, Puerto Rico. Ph.D. diss., University of Michigan, Ann Arbor.

Waide, R. B. 1991. The effect of Hurricane Hugo on bird populations in the Luquillo Experimental Forest, Puerto Rico. *Biotropica* 23:475–80.

Wai-Ping, V., and M. B. Fenton. 1988. Non-selective mating in little brown bats (*Myotis lucifugus*). *Journal of Mammalogy* 69:641–45.

Walker, L. R. 1991. Tree damage and recovery from Hurricane Hugo in Luquillo Experimental Forest, Puerto Rico. *Biotropica* 23:379–85.

Walker, L. R., D. J. Zarin, N. Fetcher, R. W. Myster, and A. H. Johnson. 1996. Ecosystem development and plant succession on landslides in the Caribbean. *Biotropica* 28:566–76.

Webb, J. P., Jr., and R. B. Loomis. 1977. Ectoparasites. Pp. 57–120 in *Biology of Bats of the New World Family Phyllostomatidae, Part II,* ed. R. J. Baker, J. K. Jones, Jr., and D. C. Carter. Special Publication 13. Museum, Texas Tech University. 364 pp.

Wetterer, A. L., M. V. Rockman, and N. B. Simmons. 2000. *Phylogeny of Phyllostomid Bats (Mammalia: Chiroptera): Data from Diverse Morphological Systems, Sex Chromosomes, and Restriction Sites.* Bulletin 248. American Museum of Natural History. 200 pp.

Whidden, H. P., and R. J. Asher. 2001. The origin of the Greater Antillean insectivores. Pp. 237–52 in *Biogeography of the West Indies: Patterns and Perspectives,* ed. C. A. Woods and F. E. Sergile. Boca Raton, Florida: CRC Press.

Whitaker, J. O., Jr. 1995. Food of the big brown bat *Eptesicus fuscus* from maternity colonies in Indiana and Illinois. *American Midland Naturalist* 134:346–60.

Whitaker, J. O., Jr., and A. Rodríguez-Durán. 1999. Seasonal variation in the diet of Mexican free-tailed bats, *Tadarida brasiliensis antillularum* (Miller) from a colony in Puerto Rico. *Caribbean Journal of Science* 35:23–28.

Whitaker, J. O., Jr., C. Neefus, and T. H. Kunz. 1996. Dietary variation in the Mexican free-tailed bat (*Tadarida brasiliensis mexicana*). *Journal of Mammalogy* 77:716–24.

Whitaker, J. O., Jr., P. Clem, and J. R. Munsee. 1991. Trophic structure of the community in the guano of the evening bat *Nycticeius humeralis* in Indiana. *American Midland Naturalist* 126:392–98.

White, J. L., and R. D. E. MacPhee. 2001. The sloths of the West Indies: A systematic and phylogenetic review. Pp. 201–35 in *Biogeography of the West Indies: Patterns and Perspectives,* ed. C. A. Woods and F. E. Sergile. Boca Raton, Florida: CRC Press.

Wiley, J. W. 2003. Habitat associations, size, stomach contents, and reproductive condition of Puerto Rican boas (*Epicrates inornatus*). *Caribbean Journal of Science* 39:189–94.

Wilkens, H., F. V. Culver, and W. F. Humphreys (eds.). 2000. *Ecosystems of the World,* vol. 30: *Subterranean Ecosystems.* Amsterdam, Netherlands: Elsevier Science.

Wilkins, K. T. 1989. *Tadarida brasiliensis. Mammalian Species* 331:1–10.

Wilkinson, G. S., and J. M. South. 2002. Life history, ecology and longevity in bats. *Aging Cell* 1:124–31.

Williams-Whitmer, L. M., and M. C. Brittingham. 1995. *A Homeowner's Guide to Northeastern Bats and Bat Problems.* University Park: Cooperative Extension, College of Agricultural Sciences, Pennsylvania State University.

Willig, M. R. 1983. *Composition, Microgeographic Variation and Sexual Dimorphism in Caatingas and Cerrado Bat Communities from Northeast Brazil.* Bulletin 23. Carnegie Museum of Natural History. 131 pp.

Willig, M. R., and A. Bauman. 1984. *Notes on Bats from the Luquillo Mountains of*

Puerto Rico. CEER-T-194. San Juan, Puerto Rico: Center for Energy and Environment Research. 12 pp.

Willig, M. R., and M. R. Gannon. 1996. Mammals. Pp. 399–431 in *A Tropical Food Web,* ed. R. B. Waide and D. P. Reagan. Chicago: University of Chicago Press.

Willig, M. R., and M. A. McGinley. 1999. Animals and disturbance. Pp. 633–57 in *Ecosystems of the World,* vol. 16: *Ecosystems of Disturbed Ground,* ed. L. R. Walker. Amsterdam, Netherlands: Elsevier Science.

Willig, M. R., and K. W. Selcer. 1989. Bat species density gradients in the New World: Statistical assessment. *Journal of Biogeography* 16:189–95.

Willig, M. R., B. D. Patterson, and R. D. Stevens. 2003. Patterns of range size, richness, and body size in the Chiroptera. Pp. 580–621 in *Bat Ecology,* ed. T. H. Kunz and M. B. Fenton. Chicago: University of Chicago Press.

Wilson, D. E. 1973. Reproduction in neotropical bats. *Periodicum Biologorum* 75:215–17.

———. Reproductive patterns. Pp. 317–78 in *Biology of Bats of the New World Family Phyllostomatidae, Part III,* ed. R. J. Baker, J. K. Jones, Jr., and D. C. Carter. Special Publication 16. Museum, Texas Tech University. 441 pp.

Wilson, D. E., and F. R. Cole. 2000. *Common Names of Mammals of the World.* Washington, D.C.: Smithsonian Institution Press.

Wilson, D. E. and D. M. Reeder (eds.). 2005. *Mammal Species of the World: A Taxonomic and Geographic Reference.* 3rd ed. Baltimore, Maryland: Johns Hopkins University Press.

Wing, E. S. 1989. Human exploitation of animal resources in the Caribbean. Pp. 137–52 in *Biogeography of the West Indies: Past, Present, and Future,* ed. C. A. Woods. Gainesville, Florida: Sandhill Crane Press.

———. 2001. Native American use of animals in the Caribbean. Pp. 481–518 in *Biogeography of the West Indies: Patterns and Perspectives,* ed. C. A. Woods and F. E. Sergile. Boca Raton, Florida: CRC Press.

Wing, E. S., C. E. Ray, and C. A. Hoffman, Jr. 1968. Vertebrate remains from Indian sites on Antigua, West Indies. *Caribbean Journal of Science* 8:123–29.

Woods, C. A. 1989. The biogeography of West Indian rodents. Pp. 741–98 in *Biogeography of the West Indies: Past, Present, and Future,* ed. C. A. Woods, ed. Gainesville, Florida: Sandhill Crane Press.

Woods, C. A., and J. F. Eisenberg. 1989. The land mammals of Madagascar and the Greater Antilles: Comparison and analysis. Pp. 799–826 in *Biogeography of the West Indies: Past, Present, and Future,* ed. C. A. Woods, ed. Gainesville, Florida: Sandhill Crane Press.

Woolbright, L. L. 1991. The impact of Hurricane Hugo on forest frogs in Puerto Rico. *Biotropica* 23:462–67.

You, C. 1991. Population dynamics of *Manilkara bidentata* (A.DC.) Cher. in the Luquillo Experimental Forest, Puerto Rico. Ph.D. diss., University of Tennessee, Knoxville.

You, C., and W. H. Petty. 1991. Effects of Hurricane Hugo on *Manilkara bidentata,* a primary tree species in the Luquillo Experimental Forest of Puerto Rico. *Biotropica* 23:400–406.

Zdzitowiecki, K., and M. A. Rutkowska. 1980. The helminthofauna of bats (Chiroptera) from Cuba: A review of cestodes and trematodes. *Acta Parasitologica Polonica* 26:187–214.

Zimmerman, J. K., M. R. Willig, L. R. Walker, and W. L. Silver. 1996. Introduction: Disturbance and Caribbean ecosystems. *Biotropica* 28:414–23.

Index